Bock

EVOLUTION, BRAIN, AND BEHAVIOR: PERSISTENT PROBLEMS

EVOLUTION, BRAIN, AND BEHAVIOR: PERSISTENT PROBLEMS

Edited by
R. B. MASTERTON
WILLIAM HODOS
HARRY JERISON

LAWRENCE ERLBAUM ASSOCIATES, PUBLISHERS
1976 Hillsdale, New Jersey

DISTRIBUTED BY THE HALSTED PRESS DIVISION OF
JOHN WILEY & SONS
New York Toronto London Sydney

Lawrence Erlbaum Associates, Inc., Publishers
62 Maria Drive
Hillsdale, New Jersey 07642

Distributed solely by Halsted Press Division
John Wiley & Sons, Inc., New York

Library of Congress Catalogue Card Number: 76-6499

Printed in the United States of America

Contents

Introduction vii

1. On the Comparative Method of Investigation
   Douglas B. Webster 1
2. Uses of the Comparative Study of Behavior
   Jack P. Hailman 13
3. Principles of the Evolution of the Brain and Behavior
   Harry J. Jerison 23
4. Rates of Evolution of the Nervous System and Behavior
   Everett C. Olson 47
5. Genetic Considerations in Studies of the Evolution of the Nervous System and Behavior
   Glayde Whitney 79
6. What Animals Should We Compare?
   C. B. G. Campbell 107
7. Grades and Clades Revisited
   Stephen Jay Gould 115
8. Interrelationships of the Major Fish Groups with Comments on Specimen Availability
   Bobb Schaeffer 123
9. The Nomenclature of Correspondence: A New Look at "Homology" and "Analogy"
   Michael T. Ghiselin 129

10. Morphological Homology and the Nervous System
C. B. G. Campbell 143

11. The Concept of Homology and the Evolution of Behavior
William Hodos 153

12. Practical Aspects of Homology Recognition
Bobb Schaeffer 169

13. In Defense of the Analog: A Commentary to N. Hotton
Stephen Jay Gould 175

14. Homology: Logic, Information, and Efficiency
Jack P. Hailman 181

15. Problems in the Comparative Study of Language
Eric H. Lenneberg 199

16. Intelligence: An Invertebrate Perspective
William J. Corning, *James A. Dyal*, and *Robert Lahue* 215

Author Index 265
Subject Index 272

Introduction

In this second volume on the evolution of the nervous system and behavior, the discussion turns from conclusions to method, from a description of what happened in history to the means of deciding what happened.

Certainly no less than in other fields, scientists trying to reconstruct the course of evolution have faced a wide variety of theoretical and practical problems. Most of these problems have been solved in due course, but many have persisted for years, some for generations, and a few since Darwin. Perhaps two of the most persistent are first, the almost cyclical reasoning underlying the comparative method itself, or the "how to" part of the science; and second, the question of homology, or how one recognizes "sameness" in different animals.

The first of these two problems stems directly from the nature of evolution. Since evolution has resulted in genetic and, usually, structural divergence, how can one deduce the neurological and behavioral characteristics of extinct animals using information that, by its nature, can only be collected from their extant (and greatly modified) progeny? Although the theoretical answer to this question is simple and direct (the characteristics of an extinct animal are most likely those characteristics shared in common by its progeny), the practical problems surrounding the application of this rule-of-thumb are neither simple nor direct, nor necessarily even solvable within the limits of present knowledge. These general issues are the subject of the first section of the volume.

The second problem, the question of homology, or the "sameness" of structure in different animals, has also persisted, although its theoretical solution seems simple too. Structures in different animals are homologous if and only if the same structure existed in their common ancestor. As examples of the use of this definition, the eyes of all vertebrates are homologous because eyes of similar structure existed in the ancestry common to them all. In contrast, the eyes of insects are not homologous to the eyes of vertebrates (even though as photoreceptors, they are analogous) because the common (wormlike?) ancestors of insects and vertebrates had no eyes whatever. But the practical problems surrounding this concept of homology can be envisioned by examining this latter argument in abstract. As it turns out, the decision that the common ancestors of insects and vertebrates were eyeless, itself depends on a previous deduction from the characteristics of the eyes of insects and vertebrates—obviously a flirtation with cyclic argument.

Nevertheless, homology remains the *sine qua non* of comparative method and the problems surrounding its use are the problems of the method itself. This issue is the subject of the second section of this volume.

Finally, the last two chapters in the volume are directed at two issues not usually nor easily integrated within discussions of the evolution of brains or behavior: first, the question of the uniqueness of symbolic language in humans; and second, the meaning for the neurosciences of recent discoveries of the behavioral abilities of invertebrates. These two problems, one at each end of the vertebrate spectrum on which the two volumes have focused, are included to serve the study of neurological and behavioral evolution more as ellipses than as periods. Certainly, the extension of the narrative of the evolution of brain and behavior in either of these directions remains an intriguing question, one which has already begun to beckon another generation of students.

BRUCE MASTERTON

EVOLUTION, BRAIN, AND BEHAVIOR: PERSISTENT PROBLEMS

1
On the Comparative Method of Investigation

Douglas B. Webster

Kresge Hearing Research Laboratory

Both scientists and laymen have been comparing animals to one another since at least the time of Aristotle, but it was not until the early nineteenth century that this pursuit was honed into a fine tool for the study of animal structure and function. Georges Cuvier (1769–1832), a French zoologist and paleontologist, analyzed the comparative method, formulated its guiding principles, and put the method systematically to work comparing structures in many vertebrates and invertebrates (see the excellent biography of Cuvier by Coleman, 1964).

His method is still used and is, in fact, so fundamental that it has passed into current thinking without notice. Yet understanding the power of this method can lead to fresh insights; understanding its mechanics can help keep current analyses as clear and orderly as were Cuvier's own.

Today the method is usually used and understood in evolutionary terms, although Cuvier, whose intellectual accomplishments predate *The Origin of Species* by 30–60 years, was an antievolutionist. He was a Rationalist above all; he looked at facts first and then devised a theory to fit those facts. Then, as now, many thinkers took the opposite approach, and pre-Darwinian evolutionary theories were inconsistent with the facts of zoology. Cuvier could only oppose the idea that species might change.

At any rate these were peripheral issues. His concern was not with a process of change but with a present reality—animal classification and what would now be called functional morphology. He explored the relationships of structure to function and of both to environment.

These relationships he called "the conditions of existence." Understanding them would be nearly tantamount to understanding the mind of the Creator. This understanding would be gained by structural and functional studies of many animals to determine how their designs fit them to their environment. He derived two rules to guide these comparative studies and the interpretation of the many facts they would reveal.

The first rule he called "the correlation of parts." It states that all parts of an organism are so intimately interdependent that they must be correlated not only with their functions but also with one another and with the animal's environment. Knowledge about one part of an animal therefore necessarily involves considerable knowledge about the whole animal. Some of these correlations have a functional basis: feathers are correlated with wings, presumably because they suit the requirements of flight better than do hair or scales. Other correlations apparently have no function: for instance, a cloven hoof is correlated with horns or antlers. Cuvier regarded the diversity of animals as examples of the Creator's "experiments," in which different means were designed to accomplish the same ends: flight, for example, was achieved by one means in birds, by another in bats. General principles of fundamental organization could be determined by comparative studies of how parts and functions are correlated in different groups.

The second rule for understanding the "conditions of existence" he called "the subordination of characters." It states that some parts of animals are so vital to survival that their possible designs are sharply limited. The common design of these least subordinate parts signals membership in one of a few large natural categories (taxa).

On both empirical and theoretical grounds Cuvier concluded that the nervous system is the least subordinate, most important part of the body, less diverse within a taxon than any other body part. External and skeletal features, however, he considered subordinate to almost every other body part; they could be designed in many different ways and therefore were not to be used to determine large classifications. This disagreed fundamentally with then current taxonomic thought. Although both schools of thought agreed that vertebrates are a natural group, for instance, Cuvier's contemporaries did so on the basis of their vertebral columns, whereas Cuvier himself pointed to their unique dorsal hollow nervous systems.

Combining these principles of the correlation of parts and the subordination of characters, Cuvier decided that there is no single *scala naturae*; instead of one basic design, modified to produce all animals, he found four different designs among metazoan animals; not surprisingly, he distinguished these primarily by the morphology of their nervous systems. Each nervous system design was the basis of a major classificatory group, or embranchement. These were:

1. the Vertebrata, characterized primarily by their spinal cord and enlarged anterior brain
2. the Mollusca, characterized by a distinct brain and scattered ganglia with nerves, but no central nerve cord
3. the Articulata, characterized by a very small brain and two distinct ventral nerve cords

4. the Radiata, the nervous system of which was based on a sparse, diffuse, radial group of nerves and ganglia in such animals as echinoderms and completely absent (!) in animals such as coelenterates.

Although he studied other parts of all these animals in order to confirm these natural groupings, Cuvier felt that all other parts were subordinate to the nervous system. This is particularly ironic because in Cuvier's time, as well as in our own, the nervous system is probably the least understood of all organ systems.

Cuvier's empirical approach and his principles of the correlation of parts and the subordination of characters were an attempt to bring reason and order into zoology. He felt that, properly used, his two principles could do for zoology what Newton's law had done for physics. Although this was optimistic, his quest for an overview was not unlike what motivated the Tallahassee conference from which these volumes grew. A group of empirical scientists, working on separate, detailed studies, came together for interdisciplinary communication with the hope that it could lead to a general understanding rather than just specific glimpses of the evolution of the vertebrate brain and behavior. In some way these paleontologists, experimental neuroanatomists, and behaviorists are Cuvier's intellectual descendents: although in these volumes it is discussed with a different vocabulary and in an evolutionary context, Cuvier's comparative method and his principles, especially that of the correlation of parts, are still used. This approach is indeed fruitful.

Until the last decade, for instance, the mammalian forebrain has been thought to have evolved quite independently of the forebrain of all other vertebrates (see, for example, Herrick, 1924; Papez, 1929; Ariëns-Kappers, Huber, & Crosby, 1936). Only mammals were thought to have visual, auditory, and somesthetic projections to the dorsal thalamus and thence to the neocortex; these sensory systems were thought to terminate at midbrain levels in nonmammalian vertebrates. The nonmammalian cerebrum was believed to be dominated by the olfactory system; small hippocampal and piriform areas received solely olfactory information; striatal structures, which somehow corresponded to mammalian basal ganglia, coordinated olfactory information and sent it to lower brain centers. Furthermore, it was believed that the nonmammalian cerebrum had only vague, indirect control of bulbar and spinal cord levels and that direct cerebrospinal pathways carrying motor information from the cerebrum to the spinal cord and bulbar nuclei were found only in mammals. These views can still be found in textbooks (e.g., Romer, 1970).

These concepts of the uniqueness of the mammalian forebrain were based on broad comparative studies of the forebrain in many vertebrates. The mammalian cerebrum is structurally unique in its characteristic six-layered neocortex that receives large and specific projections of thalamic fibers. Nonmammalian vertebrates have no six-layered neocortex, and until very recently there have been no data on projections of nonolfactory sensory modalities to their forebrain. It was therefore assumed that the neocortex had no homolog in nonmammals but had evolved *de novo* in the synapsid-mammal line and that cerebral representation of vision, audition, and touch, correlated in mammals with the neocortex, had evolved at the same time. Experimental neuroanatomy, however, particularly using degeneration techniques, has yielded results that force a reevaluation of the old ideas on the evolution of the vertebrate forebrain. Many of these results have been reported in this volume and its companion.

With these techniques, several hitherto unsuspected projections have been demonstrated in nonmammalian vertebrates. There are direct spinal cord projections to discrete nuclei in the dorsal thalamus of representative sharks, reptiles, and birds, as well as mammals. There are direct retinal projections to the dorsal thalamus in every class of vertebrates. There are auditory projections to the dorsal thalamus in all amniotes (reptiles, birds, and mammals). There are large sensory projections from the dorsal thalamus to the ipsilateral cerebral hemisphere in bony fishes, amphibians, reptiles, and birds, as well as in mammals, and to the contralateral cerebral hemisphere in sharks. The nonmammalian cerebrum is not, after all, dominated by olfactory projections; on the contrary, it has been found that olfactory information projects to only a limited portion of the cerebrum in representative sharks, bony fishes, amphibians, and reptiles, as well as mammals. Moreover, the cerebrum sends large efferent projections to lower brainstem structures not only in mammals but in sharks, teleosts, amphibians, reptiles, and birds as well; in birds, at least, some of these projections extend into the cord. These data are reviewed by Ebner (Volume 1, 1976) and Ebbesson and Northcutt (Volume 1, 1976).

It is apparent, therefore, that multiple sensory projections to the dorsal thalamus and thence to the cerebrum are not necessarily correlated with a six-layered neocortex but must be regarded as a much more general vertebrate characteristic than has previously been thought. Moreover, with the demonstration of these multiple sensory projections to the cerebrum and of motor projections from the cerebrum, the dominance of forebrain structures in the control of behavior, long established for mammals, must be at least suspected in nonmammalian vertebrates as well (Graeber, Ebbesson, & Jane, 1973).

Furthermore, because in nonmammals many specific dorsal thalamic nuclei project to structures previously regarded as striatal, one must at least entertain the idea of homologs with mammalian neocortical areas. Based on projections, for instance, it seems possible that the same fields of telencephalic cells have evolved into the teleost epistriatum, the reptilian dorsal ventricular ridge, the avian neostriatum, hyperstriatum, and ectostriatum, and, during the evolution of synapsid reptiles and mammals, into the six-layered neocortex.

There are certainly class differences in brain structures and connections but what recently emerges as even more significant is the similarity of fiber projections between parts of the brains in living vertebrates. Cuvier, looking at the nervous system in many vertebrates, observed that this system exhibited little variation from group to group, and he therefore concluded that the nervous system is of prime importance in maintaining the life of the individual and that other systems and parts are subordinate to it. Thanks to comparative studies, it is just now beginning to be realized how great is the similarity and how little the variation.

Whereas the comparative method can be fruitfully applied to broad categories, it can be even more confidently used within a restricted, closely related group. In my laboratory, I and my associates have for several years been studying the peripheral and central auditory structures of the rodent family, Heteromyidae, which includes kangaroo rats (*Dipodomys*), kangaroo mice (*Microdipodops*), pocket mice (*Perognathus*), and spiny pocket mice (*Heteromys* and *Liomys*). Thanks to the work of several paleontologists (Wood, 1935; Reeder, 1957; Shotwell, 1967) it is possible to outline phylogenies with a reasonable amount of confidence (Figure 1).

The habitats of these five genera—in the western United States through northern South America—range from tropical rain forest (*Heteromys*) to extremely arid desert (*Microdipodops*, most species of *Dipodomys*, and a few species of *Perognathus*). *Heteromys* and *Liomys*, the most tropical of the heteromyid families, have small, normal rodent middle ear cavities and quadrupedal locomotion with little if any elongation of the hindlimbs. In contrast, *Dipodomys* and *Microdipodops*, in which most or all species are arid desert dwellers, have extremely large middle ear cavities the combined volume of which exceeds that of the cranial cavity; they also have extremely long back legs, short front legs, and bipedal, ricochetal locomotion. Moreover, these characteristics are also found in African and Asian rodents that live in similar environments but that have evolved quite independently of the strictly new world heteromyids.

A quantitative morphological study of the middle ears of 26 heteromyid species shows several correlations (Webster, 1969). There is a strong relationship between tympanic membrane diameter and middle ear volume (Figure 2); in general, the dryer the habitat, the larger these measurements. The auditory ossicles and tympanic membranes have been measured and calculations made of each species' transformer ratio; this is an anatomical measure of the efficiency of the middle ear apparatus in transforming airborne sounds at the tympanic membrane to fluid vibrations within the cochlea. The transformer ratio is high for all *Perognathus, Dipodomys,* and *Microdipodops* and quite small for *Liomys*.

Initial cochlear potential studies in kangaroo rats (*Dipodomys*) showed an extreme sensitivity to low frequencies, particularly below 4000 Hz (Webster, 1962), which was greatly hampered by surgical reduction of middle ear volume. Low-frequency sensitivity in this genus has since been confirmed by electrophysiological recordings from the cochlear nuclei (Caspary, 1972; Moushegian & Rupert, 1970; Rupert & Moushegian, 1970) and by a shock-avoidance test of auditory sensitivity (Figure 3; Webster & Webster, 1972).

The adaptive value for this low-frequency sensitivity is primarily for predator avoidance. Owls and rattlesnakes, major predators of kangaroo rats, produce low-intensity, low-frequency sounds immediately prior to their strikes. Normal

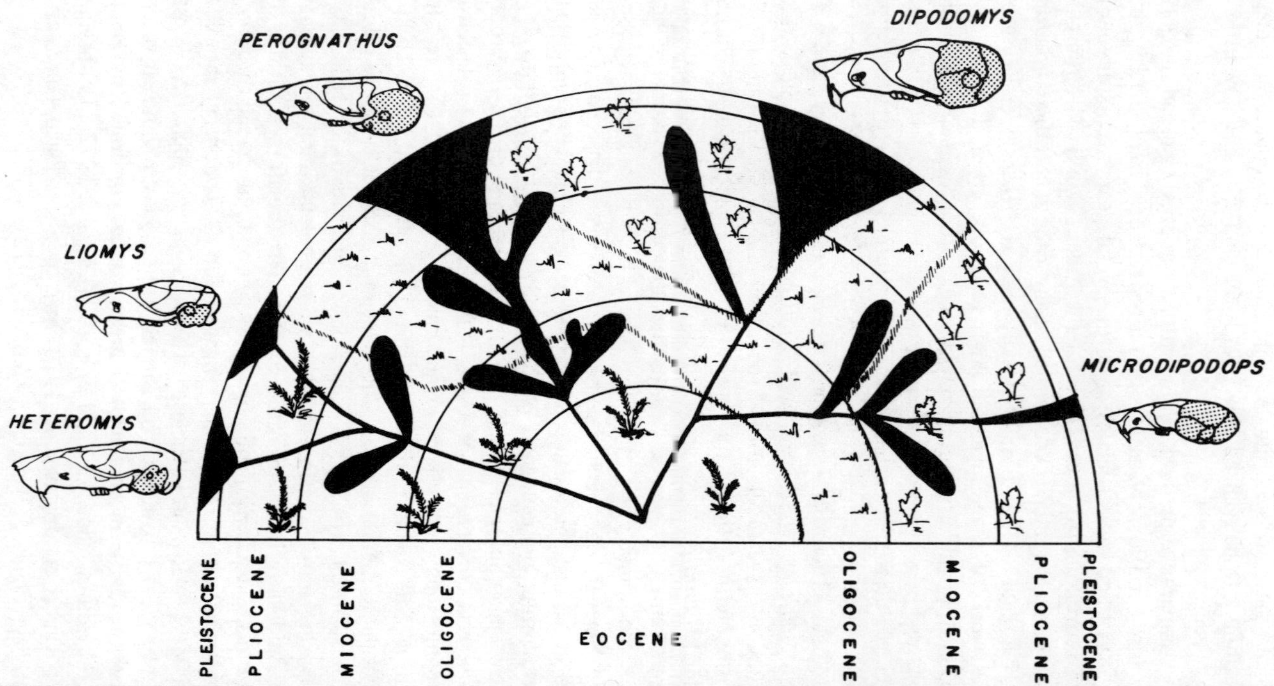

FIG. 1 Phylogenetic history of the family Heteromyidae based on paleontological studies (Reeder, 1957; Shotwell, 1967; Wood, 1935). Habitats, present and past, are indicated by diagrams of ferns for semitropical to tropical, grass for grasslands, and cacti for deserts. A representative skull of each living genus is shown with the middle ear area stippled. In both living and extinct forms the middle ears are largest in desert species.

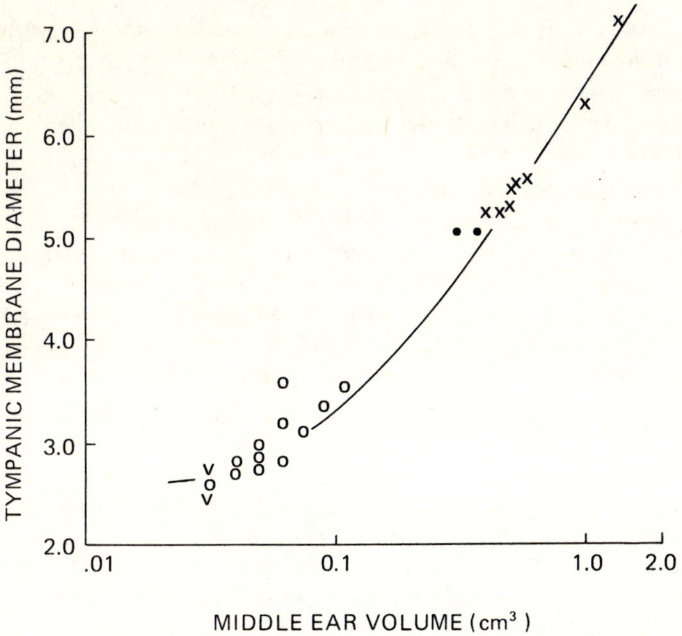

FIG. 2 The relationship of middle ear volume to tympanic membrane diameter in 26 species of heteromyid rodents (328 ears). Each point represents the average measurements for one species: (X) *Dipodomys*, (•) *Microdipodops*, (○) *Perognathus*, (v) *Liomys*. Both the middle ear volume and tympanic membrane are largest in xeric species.

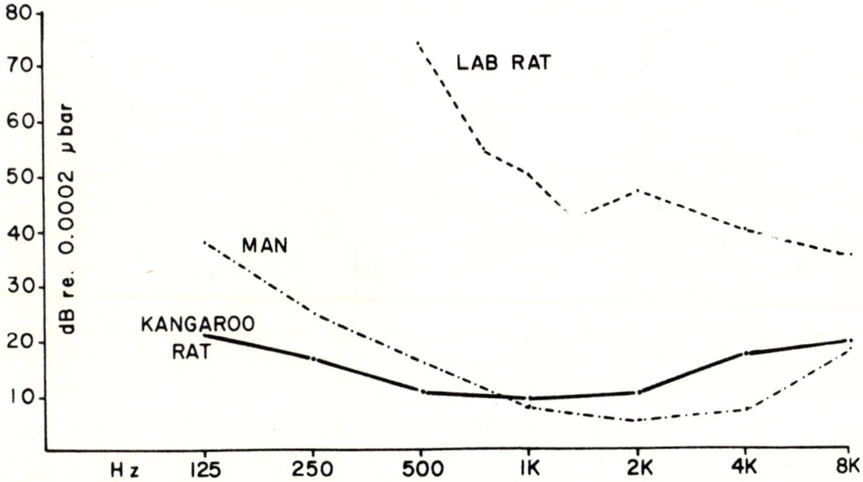

FIG. 3 Auditory thresholds of one kangaroo rat species (*Dipodomys merriami*) compared with those of man (Sivian & White, 1933) and the laboratory rat (Gourevitch, 1965) in the relatively low-frequency range from 125 Hz through 8000 Hz.

kangaroo rats usually can avoid these strikes, but kangaroo rats with reduced middle ear volumes cannot (Webster, 1962; Webster & Webster, 1971). Further studies have demonstrated that kangaroo rats can use either vision or audition to avoid sidewinder strikes, but that if both vision and audition are greatly impaired, either surgically or by environmental factors, they are highly vulnerable to predation (Webster & Webster, 1971).

From these studies it can be concluded that the large middle ear cavities, tympanic membranes, and transformer ratios of *Dipodomys* facilitate low-frequency sound reception, which is particularly useful to animals living in an arid region where there is discontinuous vegetation and a low level of background noise. Those heteromyid species living in moist, heavily vegetated areas have smaller middle ear volumes and tympanic membranes and should have more normal mammalian auditory spectra and be sensitive to high rather than low frequencies.

To test this hypothesis, the 1-μV cochlear potentials for tones from 50 to 30,000 Hz were determined in 14 species of *Dipodomys, Perognathus,* and *Microdipodops* (Figure 4). As expected, the *Dipodomys* species with the largest middle ear cavities have the most sensitive low-frequency hearing; *Microdipodops*, the relative mastoid inflation of which exceeds that of *Dipodomys* but which has somewhat smaller absolute middle ear cavities, has slightly less sensitive low-frequency hearing; some species of *Perognathus* have the poorest low-frequency hearing of these three genera (Webster & Strother, 1972). The surprising part of this comparative physiological study was that all of the species tested, regardless of middle ear volume, were more sensitive to low frequencies (below 2000 Hz) than to high frequencies (above 2000 Hz). It becomes apparent, then, that the large middle ear cavities of many heteromyid rodents do indeed facilitate low-frequency reception but that there is in addition some other factor, shared by all the test species of the family, that facilitates low-frequency reception over high-frequency reception. There is, in short, another correlation of parts not yet analyzed.

Comparative studies of the cochleae of these same heteromyid species have indicated that all the *Dipodomys, Perognathus,* and *Microdipodops* have similar modifications of the organ of Corti: an unusual thickening of the zona pectinata of the basilar membrane and highly modified border cells of Claudius and Hensen (Webster, 1961, 1968). Whereas in most mammals the cells of Hensen rest on the basilar membrane and abut against Deiters' cells, in heteromyid rodents of these genera the Claudius' cells extend inward to the cells of Deiters and support highly modified Hensen's cells. These Hensen's cells are flask shaped with large cytoplasmic processes supporting an elevated reticular lamina. They contain large amounts of nonspecific esterases–an unusual condition among mammals. It is still unknown exactly how these cochlear modifications affect sound reception, but comparative physiological and anatomical studies at my

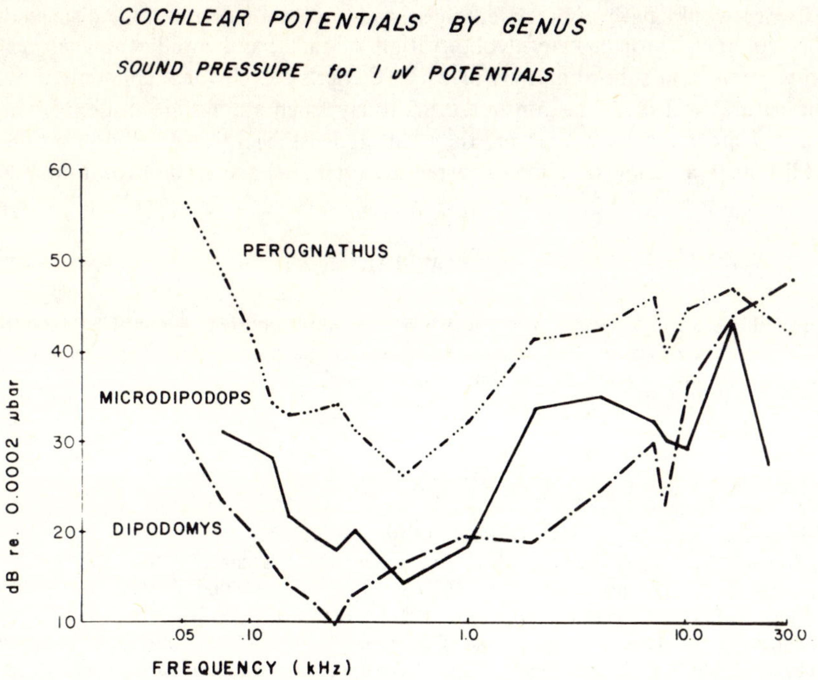

FIG. 4 The auditory sensitivity of heteromyid rodents as demonstrated by cochlear potentials (sound pressures for 1-μV potentials). The *Dipodomys* data are the mean from 22 individuals (eight species); the *Microdipodops* data from five individuals (one species); the *Perognathus* data from 21 individuals (five species). The genera with larger middle ears are most sensitive but each genus (and species) is more sensitive at low frequencies than at high frequencies.

laboratory now indicate that they must make the organ of Corti more sensitive to low-frequency sounds. This is now being investigated.

A comparative approach using anatomy and physiology in studying the auditory modifications of heteromyid rodents has therefore yielded several correlations. The enlarged middle ears and tympanic membranes, combined with large transformer ratios, facilitate low-frequency reception, which in heteromyids is made even more acute by modifications of the basilar membrane and border cells within the cochlea. This low-frequency sensitivity is particularly adaptive to animals in arid environments with discontinuous vegetation, for it helps them detect and therefore avoid the predatory strikes of snakes and owls. These correlations are apparently not limited to heteromyids: rodents of other deserts have also evolved bipedal locomotion and enlarged middle ears, and recent studies on one group of these rodents, the gerbillinae (Lay, 1972), have shown modified inner ears and extreme sensitivity to low-frequency sounds.

Cuvier would have appreciated these and the other correlations discussed in this volume. Through careful observation of facts, and intelligent analyses of correlations and subordinations of parts, Cuvier's search for a coherent view of the natural world can be continued. If, today, such studies are not expected to lead to the mind of a Creator, they can at least be looked to for continuing additions to an understanding of vertebrate brain, behavior, and evolution.

ACKNOWLEDGMENT

The author's original research reported in this paper was supported by grant NS 05800 from NIH. Ms. Molly Webster was of great assistance during preparation of this manuscript—through both intellectual discourse and writing expertise.

REFERENCES

Ariëns-Kappers, C. U., Huber, G. C., & Crosby, E. C. *The comparative anatomy of the nervous system of vertebrates including man.* New York: Macmillan, 1936.

Caspary, D. Classification of subpopulations of neurons in the cochlear nuclei of the kangaroo rat. *Experimental Neurology*, 1972, **37**, 131–151.

Coleman, W. *Georges Cuvier Zoologist*. Cambridge, Mass.: Harvard University Press, 1964.

Ebbesson, S. O. E. & Northcutt, R. G. Neurology of anamniotic vertebrates. In R. B. Masterton, M. E. Bitterman, C. B. G. Campbell & N. Hotton (Eds.), *Evolution of brain and behavior in vertebrates.* Hillsdale, New Jersey: Lawrence Erlbaum Associates, Inc. 1976.

Ebner, F. F. The forebrain of reptiles and mammals. In R. B. Masterton, M. E. Bitterman, C. B. G. Campbell & N. Hotton (Eds.), *Evolution of brain and behavior in vertebrates.* Hillsdale, New Jersey: Lawrence Erlbaum Associates, Inc. 1976.

Gourevitch, G. Auditory masking in the rat. *Journal of the Acoustical Society of America*, 1965, **37**, 439–443.

Graeber, R. C., Ebbesson, S. O. E., & Jane, J. A. Visual discrimination in sharks without optic tectum. *Science,* 1973, **180**, 413–415.

Herrick, C. J. *Neurological foundations of animal behavior.* New York: Holt, 1924.

Lay, D. M. The anatomy, physiology, functional significance, and evolution of specialized hearing organs of gerbilline rodents. *Journal of Morphology,* 1972, **138**, 41–120.

Moushegian, G., & Rupert, A. L. Response diversity of neurons in ventral cochlear nucleus of kangaroo rat to low-frequency tones. *Journal of Neurophysiology,* 1970, **33**, 351–364.

Papez, J. W. *Comparative neurology*. New York: Thomas Y. Crowell, 1929.

Reeder, W. G. "A review of Tertiary rodents of the family Heteromyidae." Ph.D. thesis, University of Michigan, Ann Arbor, Michigan, 1957.

Romer, A. S. *The vertebrate body.* (4th ed.) Philadelphia: W. B. Saunders, 1970.

Rupert, A. L., & Moushegian, G. Neuronal responses of kangaroo rat ventral cochlear nucleus to low-frequency tones. *Experimental Neurology*, 1970, **26**, 84–102.

Sivian, L. J. & White, S. D. On minimal audible sound fields. *Journal of the Acoustical Society of America,* 1933, **4**, 288–321.

Shotwell, J. A. Late Tertiary geomyoid rodents of Oregon. *Bulletin of the Museum of Natural History, University of Oregon,* 1967, **9**, 1–51.

Webster, D. B. The ear apparatus of the kangaroo rat, *Dipodomys. American Journal of Anatomy*, 1961, **108**, 123–148.

Webster, D. B. A function of the enlarged middle-ear cavities of the kangaroo rat, *Dipodomys. Physiological Zoology,* 1962, **35,** 248–255.

Webster, D. B. Comparative middle and inner ear structure of Heteromyidae. *Anatomical Record,* 1968, **160,** 447.

Webster, D. B. Comparative middle ear morphology in Heteromyidae. *Anatomical Record,* 1969, **163,** 281–282.

Webster, D. B., & Strother, W. F. Middle ear morphology and auditory sensitivity of heteromyid rodents. *American Zoologist,* 1972, **12,** 727.

Webster, D. B., & Webster, M. Adaptive value of hearing and vision in kangaroo rat predator avoidance. *Brain, Behavior, & Evolution,* 1971, **4,** 310–322.

Webster, D. B., & Webster, M. Kangaroo rat auditory thresholds before and after middle ear reduction. *Brain, Behavior, & Evolution,* 1972, **5,** 41–53.

Wood, A. E. Evolution and relationships of the heteromyid rodents. *Annals of the Carnegie Museum,* 1935, **24,** 73–262.

2

Uses of the Comparative Study of Behavior

Jack P. Hailman

University of Wisconsin

Zoologists often refer to "the" comparative method, but it seems doubtful that all comparative studies have exactly the same aims. The ethologist studies a range of species in order to learn various things relating to the determinants of behavior. To understand the ethologist's comparative method, it is necessary first to review these behavioral determinants.

CAUSES AND ORIGINS OF BEHAVIOR

The question "why does an animal do that?" may be several questions simultaneously. Suppose "doing that" is the micturation pattern of adult male dogs that lift one hind leg next to a vertical object on which they urinate. The question "why does the dog do that," may then be, first, what the stimuli are that elicit the micturation, or the muscle system underlying the movements involved, or the central nervous mechanisms controlling the behavior. Alternatively, it may be what events in the development of males cause them to differ from females in micturation patterns, because puppies of both sexes do not differ in such behavior; or, it may be what function such behavior plays in the behavior of the animals, because it is believed that the difference between males and females is adaptive in the evolutionary sense. Finally, the question may be how such behavior has evolved during the past history of phylogeny. In order to understand micturation behavior in any complete sense of "understand" all these questions about control, ontogeny, function, and phylogeny of behavior must be answered.

Dynamic Control

Behavior of any sort may be conceived of as the transduction by an organism of high informational inputs having relatively low energy to low informational outputs having relatively high energy. This cybernetic statement of behavior differs in subtle ways from the more usual "stimulus–response" paradigm, although a discussion of this point is out of place here. The focus here is on understanding these dynamic transformations of information as they come from the environment and pass through the organism. Sometimes the dynamic control of behavior is studied merely at the behavioral level itself: analysis of stimuli, motivational factors, and observable motor responses. When more detail is sought, the study becomes one of neurological, endocrinological, and myological mechanisms of behavior. All these factors may collectively be referred to as "the dynamic control of behavior," a concept similar to what the British call "causation."

Therefore, returning to the initial example, an answer to "why does the dog do that?" may be "because he smells a particular olfactory stimulus." This answer, of course, is only one of many, as it merely tells the immediate environmental stimulus that elicits the behavior. The neurophysiological basis of the behavior, with its attendant complexities, can also be questioned. In either case, the immediate dynamic control of micturation in a single dog is being sought.

Ontogenetic Development

Frequently, the kind of dynamic control shown in animal behavior is strongly influenced by the past history of the individual animal. Indeed, all behavior has an ontogenetic history of continued interaction between the behaving organism and its environment. Some behavior is relatively "environmentally resistent," usually reaching about the same phenotypic end point regardless of the animal's past history and experiences. Other behavior is very dependent on the past histories and experiences of the animal. In either case, the genetic endowment of the animal relative to the behavior of interest and the factors in the environment that shape how that endowment shall be expressed in terms of dynamic control of behavior in the adult can be sought. Both the dynamic control and the ontogenetic development of behavior are determinants that work at the level of the individual organism; that is, these determinants may be studied in single individuals.

Returning again to our initial example, the question "why does the dog do that?" can be answered differently this time. Experiments can be cited showing that casterated puppies never respond to odors by urinating, whereas adult males that have developed the behavior continue to show it after castration (Berg, 1944). In other words, events in the past history of the individual's life affect the individual's present behavior. Not only the capacity to respond to certain stimuli but also (in this example) the very motor pattern used is affected, because puppies, like adult females, squat and only the adult males lift their legs during urination. Naturally, the antecedent events in the ontogeny of behavior can be very complex events of learning and experience, as opposed to relatively simple events of this example that are caused by the onset of hormonal flow.

Biological Function (Selective Advantage)

The genetic endowment of an individual is the product of the forces of natural selection acting on the variation in a population of which the individual is a member. Elsewhere in this volume (Chapter 14) I discuss briefly the misunderstandings incurred by the word "function" in behavior. To the ethologist, the biological function of a behavioral pattern is a loose statement of the forces of natural selection that are acting on the phenotypic variation in the population. One "function" of eggshell removal in gulls is loosely and teleologically spoken of as being a defence against discovery of the nest by predators (Tinbergen, Broekhuysen, Feekes, Houghton, Kruuk, & Szulc, 1962). This term "function" is merely shorthand for saying that individuals failing to remove eggshells suffer great nest predation and hence leave few offspring; therefore their genes are trimmed from the population by natural selection.

Returning once again to the initial example, the question "why does the dog do that" can be answered in still another way. The necessity to life of maintaining certain chemical balances in the body necessary for proper metabolism may be cited and how kidney function and elimination of nitrogenous wastes contributes to such critical balances discussed. For male dogs, of course, this is only part of the function of the behavior, for it fails to explain why dogs "do that" so frequently, why they respond to odors that do not have such effects in females, and why they use a different motor pattern of micturation. Although the differences in both endocrinology and ontogeny that produce the observed differences between male and female may already be known, it is still necessary to understand the selective function of the male's unique behavior. Therefore, the advantages of territories and spacing apart of males and how males communicate with one another by olfactory stimuli must be discussed. This line of inquiry leads to a discussion of the accuracy by which odiferous urine can be deposited on a vertical surface by different motor patterns of micturation, and so on. All such answers are attempts to summarize why the observed behavior is favored by natural selection over its alternatives.

Phylogenetic History

The process of speciation—whereby an ancestral species splits into two noninterbreeding populations—combined with the forces of natural selection through time have produced the phylogenetic history of behavioral patterns of present-day species. These histories are difficult to reconstruct from the fragmentary nature of the evidence available but are no less real for it. Elsewhere in this volume (Chapter 14) I discuss how behavioral patterns having the same function differ (leaf-turning behavior of forest floor birds). Why they differ is a question that can only be answered phylogenetically. The brown thrasher (*Toxostoma rufum*), having evolved from an ancestor with long, thin body proportions, necessarily stands generally parallel with the ground. Its beak is close to the ground in a position to turn leaves aside, whereas its feet are positioned far to the rear. In contrast, the superficially similar fox sparrow (*Passerella iliaca*), having evolved from an ancestor of shorter and squatter proportions, stands more nearly vertical with its head high above the ground and nearly over its feet. It is not surprising that these two species, having rather different ancestral materials for natural selection to shape further, have evolved quite different but functionally equivalent methods for turning leaves: the thrasher flicks them aside with the bill, the sparrow scratches them behind with the feet.

Certain aspects of the question "why does the dog to that" may even require such historical answers. For instance, it may be inquired why the dog deposits scent with its urine instead of by secretions from suborbital glands or other special tissues on the surface of its body. The answers to such questions may be difficult, indeed, to obtain, for it is not obvious how canine ancestors have been preadapted toward developing this particular kind of olfactory communication rather than some other kind. Perhaps, for instance, canines (being very widely ranging animals) may have been adapted to frequent elimination of wastes so as to carry no more body weight than necessary. This habit could have preadapted ancestors of dogs to using the frequent micturation patterns for the additional function of olfactory marking. This preadaptation could be contrasted, for instance, with the behavior of ancestral browsing deer, the heads and bodies of which were frequently in contact with vegetation. Their behavior may preadapt them toward the evolution of skin glands to be rubbed against trees as a method of scent marking. The evolutionary explanation here is speculation created for the purpose of completing the example raised at the outset but it is no less important for that.

Overview

Four completely separate classes of answers to the question "why does the dog do that" have now been identified. It is my thesis that a behavioral pattern is not understood in any complete sense until these four classes of answers, which are not contradictory but, instead serve to reinforce one another in a total explanation of behavior, can be provided. It is irrelevant whether the initial question is treated as an invalid and naive one that asks many questions at once, or whether it is viewed as the ultimate and sophisticated question to be answered. The important point is that behavior has not a single determinant but four separate and interrelated classes of determinants.

In sum, there are two immediate "causes" of behavior (control in the individual and maintenance by natural selection in the population) and two historical "origins" of behavior (ontogenetic development in the individual and phylogenetic evolution in the population). These four "causes and origins" or determinants of behavior are summarized in Table 1. The table schematizes the inherently interdisciplinary nature of ethological research: the ethologist must be a bit of a physiologist to understand control, a bit of a developmental geneticist to study ontogeny, a bit of an ecologist to study adaptive function, and a bit of an evolutionary systematist to study phylogenetic origins. It is likely that the ethologist never fully satisfies any of the workers in the parental fields between which he attempts to construct bridges; yet he tries, and one of his most powerful methods is the comparative study of behavior.

THE COMPARATIVE METHOD IN ETHOLOGY

The study of similar behavior in a host of species may reveal various aspects of the behavioral determinants enumerated above. Ideally, the choice of species for study includes both animals taxonomically related and unrelated and animals with similar and dissimilar ecologies. Only with such a conscious parametric design are the proper controls present for evaluating the similarities found in the behavior of the species studied. The design is schematized in Table 2, where the cells are lettered for reference below. Depending on the results of the comparative survey, certain clues as to the determinants of the behavior under study emerge.

Phylogeny

If the behavior (or morphology, for that matter) is similar for the species in Cells A and B but different in species of Cells C or D of Table 2, then the behavior correlates with the evolutionary taxonomy of the animals and not with the ecology of the species involved. For instance, my own comparative survey of the

TABLE 1
Determinants of Behavior

Type of determinant	Level of Action	
	Individual	Population
Causes	Dynamic control	Adaptive function
Origins	Ontogenetic development	Phylogenetic origin

TABLE 2
Ideal Design of Comparative Studies[a]

Taxonomic relationships	Ecological situations: Similar	Dissimilar
Close	A	B
Distant	C	D

[a] All four cells should be filled with sufficient sample sizes of species studied.

oil droplets in the cone photoreceptors of anuran amphibians reveals a correlation between the presence of these subcellular structures and the taxonomy within the order Anura (Hailman, in press). All six suborders and 16 of the 17 families were surveyed in a study of 97 species, and oil droplets were found in only two of the suborders. Within these suborders species with tremendously variable ecologies were shown to have the oil droplets, whereas species in the other four suborders always lacked the droplets, even though such species show the same broad diversity of ecological types. Because the two suborders possessing oil droplets are not considered to be particularly closely related, and because details of the structures differ in the two suborders, it may be assumed that the best evidence is that cone oil droplets have arisen twice in independent lines of anuran ancestors.

A similar comparative survey (Jaeger & Hailman, 1973) of phototactic behavior in 121 species of anurans has failed to reveal clear taxonomic correlations as the morphological study. Certain similarities of behavior occur in taxonomic groupings (e.g., bufonid toads are nearly always photopositive, whereas arrow-poison frogs and their relatives in the Dendrobatidae are modal with a peak intensity preference in phototactic tests), but in general taxonomy is not a strong predictor of phototactic behavior.

Taxonomy and Comparative Behavior

Many persons viewing ethology from afar receive the impression that comparative studies of behavior are primarily motivated by an interest in classifying the organisms of study. Although this motive undoubtedly impels some, taxonomic results are often more a byproduct of such studies than a focus. The use of the comparative method in this way does raise an interesting epistemological question: that of possible circularity of argument. For instance, if fundamental correlations (Table 2) require judgements of taxonomic relatedness of the species studied, how can the behavior observed be used as a taxonomic character to classify the animals? The reply to this possible criticism, I believe, lies in a proper understanding of the multiplicity of characters on which phylogenetic taxonomy is based. There are few, if any, classification schemes of animals built solely on behavior; usually behavior merely confirms taxonomies established on primarily morphological grounds. A classic ethological example involves the taxonomy of waterfowl, in which certain behavioral patterns have been discovered to parallel the existing classification (Delacour & Mayr, 1945). The inherent danger of circularity in tracing the phylogenetic history of a behavioral pattern by taxonomic correlation when the taxonomy itself is built on comparative behavioral studies is virtually nonexistant. I know of no example in which this possible complication is actually a factor.

Function

If the behavior (or morphology) is similar among the species occupying Cells A and C but different in species of Cells B and D of Table 2, then the comparative survey indicates something about the adaptive function of the character under scrutiny. For instance, the phototactic behavior in anuran amphibians noted above (Jaeger & Hailman, 1973) shows such a correlation. Species of different suborders that show photonegative tactic behavior share a common environmental factor. Some are active only on the darkest moonless nights, others are active at night in environments with heavily forested canopy, and one species even lives in termite mounds in Africa; all share the factor of being active under extremely low levels of illumination, even if rhythmically they are not classifiable as "nocturnal."

An example from another group of animals is mentioned by Hodos (Chapter 11) in which cliff-nesting kittiwake gulls differ in many respects from their ground-nesting relatives (E. Cullen, 1957). This difference in itself is insufficient to support the notion that the peculiar behavioral patterns of the kittiwake are functional adaptations to the cliff-nesting environment, for the comparison is essentially between Cells A and B in Table 2. However, the cliff-nesting swallow-tailed gull of the Galapagos islands, more distantly related, shows many of the same traits as the kittiwake (Hailman, 1965). Indeed, in some cases, alternative behavioral solutions are used to solve the same problem: e.g., the kittiwake builds an unusually deeply cupped nest, which presumably functions to prevent eggs from rolling out of the nest and over the cliff edge, whereas the swallow-tailed gull builds its nest of rocks and sea urchin spines instead of vegetation, resulting in a nest from which it is equally difficult to roll eggs.

Cell C of Table 2 is filled even more solidly, however, by cliff-nesting terns, in the same order but in a different taxonomic family than the gulls (J. M. Cullen & Ashmole, 1963). Not only the terns but also the quite distantly related cliff-nesting gannets of a different avian order, show behavior essentially similar to the kittiwake's (Nelson, 1967). Therefore, the similarity in species of Cells A and

C and the difference between these and the behavior seen in species that fall in Cells B and D of Table 2 lends strong evidence that the behavior is adaptive to cliff-nesting conditions.

It should be emphasized that in the case of inferring adaptive functions of behavior comparatively, as in inferences of phylogenetic origins above, the hypotheses erected are testable. If the comparative method in these cases is valid, it should predict the behavior of yet unstudied species. In fact, the swallow-tailed gull behavior (Hailman, 1965) was explicitly studied as a test of Cullen's (1957) functional interpretation of kittiwake behavior. Implicitly, this kind of testing by predicting behavior from either the ecology or the taxonomy of unstudied species goes on continuously in ethology much as new fossil finds confirm reconstructions of missing parts of animals or phylogenies erected on previous evidence. Those who assert that phylogenetic hypotheses are inherently untestable carry the burden of showing how comparative study of new forms fails to meet logical criteria of valid empirical science.

Other Uses of the Comparative Method

The examples cited above have two characteristics that cannot always be found in animals deserving study: they are of relatively close relationship (within a single class of animals) and there are an abundance of species available for comparative purposes. Clearly, the comparative method in its ideal form (Table 2) has limitations if it is desired to trace phylogenetic patterns among all vertebrates or to investigate adaptive patterns in monotreme mammals. In parallel with laboratory experimental methods, the comparative method increases in value with its sample size, i.e., the number of species being compared. When species are few and their phylogenetic relationships are clouded in the distant past of unfossilized ancestors, the comparative method reaps fewer conclusions of trust.

However, with these things said, it is still useful to study a diversity of animals. Even when the courses of phylogeny are unknown and the relationships of extant forms unclear, interesting patterns can be detected in comparisons. Sometimes these patterns can be arranged in a series of increasing complexity, or other dimension. The ordered series is one paleontologists refer to as "grades," and although the evidence may be inadequate to show such steps on a continuum to be likely phylogenetic "clades"—or may even show them conclusively not to be—the gradational stage still provides two interesting inferences. First, if behavior and structure are correlated in such a series, the comparison provides strong clues to the control mechanisms underlying the behavior and thus contributes to another of the behavioral determinants (Table 1). Second, the series, although not phylogenetic, may still provide clues to the kinds of stages that may have occurred in true phylogenetic lines.

Comparative study is useful even if a "pseudoevolutionary," gradational series fails to emerge from the comparisons. At the least, survey of behavior or morphology in diverse organisms can often uncover species particularly suitable for study of physiological mechanisms or ontogenetic development (see Table 1). Therefore, although the chief thrust of comparative study is on function and phylogeny, it is also useful in understanding other determinants of behavior. Furthermore, comparative study may uncover qualitative differences in some behavioral process, even if the determinants underlying such differences are initially unclear. For instance, the comparative learning studies by Bitterman (Volume I, Chapter 11) have uncovered distinct learning strategies in goldfish, pigeons, and certain mammals. The adaptive function of these different patterns is not clear, nor have sufficient species yet been studied to indicate whether there exists a "fish," "bird," and "mammal" kind of learning. Despite the fact that nothing can be said of the evolution and function of these learning styles, however, the differences uncovered by comparative study are real and open new doors for future investigation. They may yield real clues to adaptiveness or evolution on further comparative survey or they may uncover structure–behavior correlations when further work is performed.

CONCLUSIONS

One point seems inescapable: comparative study provides a method for getting at the adaptive function and phylogenetic history of behavior, determinants difficult to study by any other method. Comparative study is also useful in selecting species for ontogenetic and physiological studies, as well as in providing valuable series of "grades" that suggest possible evolutionary sequences of a general nature and in discovering new qualitative differences in behavior among animals. Periodically, comparative psychology appears to suffer a soul searching, in which someone laments the paucity of animal species studied and painfully documents the tremendous numbers of studies on rhesus monkeys, white rats, white Carneau pigeons, and a few other selected types. The trouble today, however, may lie more with editors than authors. Papers on diverse organisms issue from psychology departments into such journals as *Behaviour* and *Animal Behaviour*, whereas traditional psychological journals continue to devote their pages to the domesticated albino misfits of man's kingdom.

REFERENCES

Berg, I. A. Development of behavior: the micturition pattern in the dog. *Journal of Experimental Psychology,* 1944, **34,** 343–368.

Bitterman, M. E. Issues in the comparative psychology of learning. In R. B. Masterton (Ed.), *Evolution of brain and behavior in vertebrates.* Hillsdale, New Jersey: Lawrence Erlbaum Assoc., 1976. Chapter 11, Issues in the comparative Psychology of Learning.

Cullen, E. Adaptations in the kittiwake to cliff-nesting. *Ibis,* 1957, **99,** 275–302.

Cullen, J. M., & Ashmole, N. P. The black noddy *Anous tenuirostris* on Ascension Island: Part 2, Behaviour. *Ibis,* 1963, **103b,** 423–446.

Delacour, J., & Mayr, E. The family Anatidae. *Wilson Bulletin,* 1945, **57**, 3–55.

Hailman, J. P. Cliff-nesting adaptations of the Galapagos swallow-tailed gull. *Wilson Bulletin,* 1965, 77, 346–362.

Hailman, J. P. Oildroplets in the eyes of adult anuran amphibians: A comparative survey. *Journal of Morphology*, in press.

Jaeger, R. G., & Hailman, J. P. Effects of intensity on the phototactic responses of anuran amphibians: a comparative study. *Zeitschrift für Tierpsychologie,* 1973, 33, 352-407.

Nelson, J. B. Colonial and cliff nesting in the gannet. *Ardea,* 1967, **55**, 60–90.

Tinbergen, N., Broekhuysen, G. J., Feekes, F., Houghton, J. C. W., Kruuk, H., & Szulc, E. Egg shell removal by the Black-headed Gull, *Larus ridibundus* L.: a behaviour component of camouflage. *Behaviour,* 1962, **19**, 74–118.

3

Principles of the Evolution of the Brain and Behavior

Harry J. Jerison

University of California, Los Angeles

INTRODUCTION

Few principles of brain function relevant for evolutionary analyses have been developed within an evolutionary framework, and some were actually discovered before Darwin's great synthesis of evolutionary theory. Early in the nineteenth century some of the basic correlations between structure and function in nervous systems were beginning to be appreciated from the discovery of the sensory role of the dorsal horn and the motor role of the ventral horn of the spinal cord. At the level of the brain, the doctrine of localization of function was proposed by Broca and the phrenologists and opposed by Flourens' version of the mass action hypothesis (see Brazier, 1959). To the extent that differentiation of functions could be associated with morphological differences it was possible to develop both comparative neuroanatomy and paleoneurology within an evolutionary framework. Differences among the brains of fossil animals could be related to evolutionary change.

Other principles, developed independently of evolutionary points of view and after the publication of *The Origin of Species*, included the neuron doctrine, which established a possible basic unit of analysis of neural activity (see Brazier, 1959). Recent work on "miniature nervous systems" may have found an even more appropriate unit for analysis of neural activity as it occurs in the brain (Kennedy, 1967), one that can serve in the role of cell assemblies as discussed by Hebb (1949). In addition, there are now modern statements of doctrines of localization of function (Woolsey, 1958) and of mass action (Lashley, 1950; John, 1972). Less widely appreciated, but also important for evolutionary analysis, are the quantitative studies of relations between microscopic and gross structures in the brains of different animals (Bok, 1959; Harman, 1957; Stephan, Bauchot, & Andy, 1970).

The only basically evolutionary principle of neural activity is that of encephalization or corticalization of function, which states that as advances on the phylogenetic scale have taken place more rostral parts of the nervous system have taken over specific functions. This principle is certainly wrong unless a vaguely teleological definition of "function" is accepted, and even then its validity is uncertain (Weiskrantz, 1961).

The direct analysis of the evolution of the brain from the data of fossil endocasts yielded a number of important structural principles, the most important of which appeared a few years after Darwin's publication. This was Lartet's (1868) and Marsh's (1874) discoveries that relative brain size increased with the passage of geological time in at least some lineages of the mammals. This is a principle of structural encephalization.

LOCALIZATION OF FUNCTION AND THE PRINCIPLE OF PROPER MASS

According to the localization doctrine, neuropsychological functions are controlled or determined by localized structures in the nervous system. For example, if a localized region of a mammal's retina is stimulated with light, neural units respond in localized regions of the retina, optic nerve, optic tract, superior colliculi, lateral geniculate bodies, and a specifiable region of the cerebral cortex. The procedures for determining the limits of this projection system are complex and the specification is itself far from complete at this time. For example, a portion of the cortex may be characterized as "visual cortex" by the latency of responses, by anatomical studies of pathways between thalamus and cortex, and by other methods, while it is recognized that neural activity in other regions of the brain may also be altered by the presentation of a simple stimulus to the peripheral sense cells of the retina. As a first approximation it is appropriate to think of a restricted region of the cortex as receiving the primary projections from the retina, and this is an instance of "cortical localization."

In the analysis of the evolution of the brain and its role in the adaptive radiation of animals, the doctrine of localization is the basis for the important "principle of proper mass":

> The mass of neural tissue controlling a particular activity is appropriate to the amount of information processing involved in performing the activity.

This principle implies that in comparisons among species the importance of an activity in the life of each species is reflected by the absolute amount of neural tissue for that activity. It also implies that, within a species, the relative masses of neural tissue associated with different behaviors are related to the relative importance of the behaviors in the species. Among the mammals, "visual" species have enlarged superior colliculi and visual cortex, and "auditory" species have enlarged inferior colliculi and auditory cortex.

An example, presented in more detail below, illustrates the application of the principle. It is often stated, probably with some justice, that primitive mammals tend to have enlarged olfactory bulbs and therefore have been "olfactory" animals. One proposed measure of the progressiveness of a species of mammals is whether there has been replacement of olfaction by other senses and relative reduction of the size of the neural olfactory system (e.g., Tilney, 1931). A quantitative analysis reveals, however, that as the mammals have evolved the absolute size of the olfactory bulbs in many species has actually increased. It is only when the olfactory bulbs are compared with other parts of the same brain that they may appear relatively smaller. It may therefore be conjectured that many living mammals are more efficient in their ability to use olfactory information, or in the amount of olfactory information that they use, compared to their fossil ancestors. The elaboration of other systems in the brain has been so much greater in the evolution of "progressive" mammals, however, that their olfactory system has been overshadowed in size.

The doctrine presented in opposition to localization of function is that of mass action, generally associated with Lashley's early work (1929). As currently conceived, that doctrine can be stated somewhat as follows: complex behavioral functions tend to be governed by an extensive and diffuse network of neural structures. The degree of incapacitation with respect to those functions following brain damage, therefore, tends to be related to the amount of tissue destroyed, more or less independently of the locus of the destroyed tissue. It was probably because of the truth of this principle, at least as stated here, that Lashley was able to recognize the role of gross brain size as an anatomical correlate of intelligence of "behavioral capacity" (Lashley, 1949).

NEURONS AND MINIATURE NERVOUS SYSTEMS

The evolutionary significance of the neuron arises from its near universality as an information processing unit in the metazoan nervous system. The same near universality is probably true of the "miniature nervous system" considered as a unit, although such systems have only recently come under intensive study. Their widespread occurrence means that a fundamental evolutionary adaptation must have occurred before the major multicellular phyla became differentiated from one another, because miniature systems have been found in many species from many different phyla. These have included mollucsks (Kandel, 1967), arthropods (Kennedy, 1967; Wilson, 1968), and vertebrates (Rovainen, 1967). At present it seems likely that the miniature system should be regarded as the fundamental unit or building block with which more complex or elaborate neural systems are constructed. Its role may be as a "prewired cell assembly," in Hebb's sense (1949), in which specific nerve cells act together to carry out fairly elaborate actions. The point is important, because it suggests that some apparently innovative evolutionary "experimentation" may actually represent the persistence of very ancient behavior mechanisms inherent in the structure of synaptic nervous systems, because comparable behavior occurs in miniature nervous systems.

Behavioral functions that occur in miniature nervous systems cannot be considered as newly evolved functions in the vertebrates. For this reason, simple sensory-motor integrations must be considered as a common feature of the nervous systems of most multicellular animals and of all vertebrate species. The evidence with respect to miniature nervous systems also indicates that simple kinds of learning, habituation in particular but perhaps also conditioning, occur in these systems (see Bullock, 1967). In other words, learning as well as "lower" functions may actually be a property of the elementary units (miniature systems) that go into the building of a vertebrate brain.

ENCEPHALIZATION AND CORTICALIZATION OF FUNCTION

Encephalization, as indicated above, is the major evolutionary principle that has been proposed for comparative neurology. It is based on several kinds of evidence, in particular the effects of brain damage in different species at different levels of the nervous system. For example, for many years it was considered that lesions of the visual cortex in man produced complete blindness, that comparable lesions in monkeys produced almost complete blindness but the capacity to discriminate luminous flux remained, whereas in rats and cats they produced loss of habits based on brightness discrimination, but those habits could be relearned, indicating that the visual capacity was not completely lost. Weiskrantz (1961) rejected the interpretation of the data as evidence of encephalization (corticalization) and reached essentially the conclusion presented in the next paragraphs. Encephalization of somatosensory functions (Marquis, 1934), of motor functions (Ruch, 1935), of auditory functions (Ades, 1959), and of primate functions as a whole (Noback & Moskowitz, 1962) have also been described. All of these analyses are subject to the kind of criticism presented by Weiskrantz.

As an evolutionary argument the preceding description of encephalization can almost be dismissed out of hand. The facts imply no more than that several species of mammals have evolved different sets of functions (in the strict sense: input–output relationships) for processing visual information and that damage to cortical and subcortical brain structures has different effects in these species. There is no evidence of an evolutionary succession, and there is little reason to consider rats, cats, and monkeys as more progressive or less progressive than one another, or as representing any kind of phylogenetic series, although they may represent "grades" of encephalization (cf. Gould, Chapter 13).

On the basis of the other principles discussed here, a simple statement of encephalization or corticalization of function makes little sense. It is much more likely that nervous tissue is conservatively organized in the sense that a particular function (input-output relationship) once established in particular networks is performed by homologous networks in descendant species in an evolutionary series. This is the evolutionary "conservation" principle (Stebbins, 1969) discussed below. It would be an unnecessary burden for the evolutionary process to evolve new systems to solve a problem already solved by an existing system. The doctrine of corticalization would require that a function established in, e.g., a midbrain network in a primitive species, be transferred to a cortical network in the descendant species. Furthermore, this kind of transfer would have to have occurred independently for many functions in many different evolutionary lineages. Encephalization is, therefore, an unlikely physiological "fact," and it is even more far fetched to think that it became a common process in different genetic systems evolving at the same time. It is much more likely and makes good evolutionary sense that sets of related visual functions were elaborated differently in different species when they entered their varied niches, for example, that form vision is different in rat, cat, and monkey and is therefore differently organized, rather than being the same form vision handled by different neural systems.

A comparative sketch of the auditory system may clarify this point. Auditory information used by early land vertebrates was probably analyzed by a very small number of neural elements. Peripheral auditory systems near the sense cells and in the medulla of living amphibians and reptiles have only a few hundred or perhaps a few thousand neurons at most (Wever, 1965). Although little is known about the auditory worlds of those lower vertebrates, it may be assumed that the level of auditory analysis cannot have been very sophisticated at any time in their evolutionary history.

With the evolution of the much more elaborate auditory apparatus in mammals, more sophisticated auditory information may be used, such as information for echolocation by bats, in which tectal (midbrain) systems are emphasized (Grinnell, 1970). If all auditory functions are grouped together as an imprecisely defined auditory "function," it can be correct to state that "auditory function" is much more disrupted in bats than in lizards as a result of lesions in the midbrain. Audition is clearly more encephalized in bats, but only because of the broad use of the term "audition" to cover the full range of functions in which the auditory system is involved. It makes better sense, however, to recognize that in the bat there are functions of hearing that do not exist in the lizard, and that those functions are localized in more rostral structures than are most of the functions that are common to bat and lizard.

A restricted view of encephalization could refer to the likelihood that the evolution of new functions involved the elaboration of new neural structures, which tended to be more rostral in the central nervous system. This view, too, is questionable. The elaboration of such hindbrain structures as the cerebellum, for example, was highly correlated with the elaboration of the forebrain in mammals. The most nearly correct statement would probably be that the functions that first appeared in the evolution of mammals and birds were governed by forebrain structures more than other parts of the brain. These were, generally, sensory-motor functions and coordinating or integrative functions related to information processed by distance-sense modalities. As a result, the evolution of the brain in birds and mammals has generally been characterized by greater enlargement of the forebrain than of the rest of the brain (save possibly the cerebellum in some species).

STRUCTURAL ENCEPHALIZATION: THE BRAIN AS A WHOLE

Structural or morphological encephalization did occur in the evolution of the vertebrates. The evidence can be developed from the fossil record by the analysis of relative brain size. There are some surprises in the data. For example, there is fossil evidence that encephalization has been minor in the evolution of reptiles and clear evidence of significant encephalization in the very beginning of the evolution of birds and mammals from their reptilian ancestors. The broad trend is mapped by a series of convex polygons of the type shown in Figure 1. (The living mammals are represented by the polygon outlined by the broken line at the bottom; the smaller mammalian polygon is based on an extensive sample of living mammals [Crile & Quiring, 1940] in which insectivores are poorly represented. The extended mammalian polygon includes insectivores described by Bauchot & Stephan, 1966. See Jerison, 1969, 1971b, 1973, for additional information.)

Although not included in Figure 1, birds form a population that overlaps the mammalian polygons and does not overlap the reptilian one. The "archaic ungulates" in Figure 1 are indicated by the numbers 1–14. (The numbers 15–18 refer to "archaic carnivores" of the order Creodonta. The particular species of creodonts in Figure 1 are middle and upper Eocene and lower Oligocene, that is of from about 30 to 50 million years ago, and are considered only in passing here. I use the term "archaic" to refer to an order that has become extinct in its adapative zone and has apparently been replaced in that zone by species of a "progressive" order.) The letters T, P, E, and D refer, in order, to the Jurassic mammal *Triconodon*; the Eocene multituberculate, *Ptilodus*; and the living hedgehog and opossum, *Erinaceus europaeus* and *Didelphis marsupialis. Triconodon* provides evidence of the status of the oldest mammal with known endocast–about 150 million years old. The other three represent other forms that are similar to Mesozoic mammals in body size and shape. They may be thought of as representing the status of relative brain size that is appropriate for life in mammalian Mesozoic niches.

The interesting feature of Figure 1 is that the archaic mammalian polygon (stippled), which contains the species from "Mesozoic niches," lies at or slightly

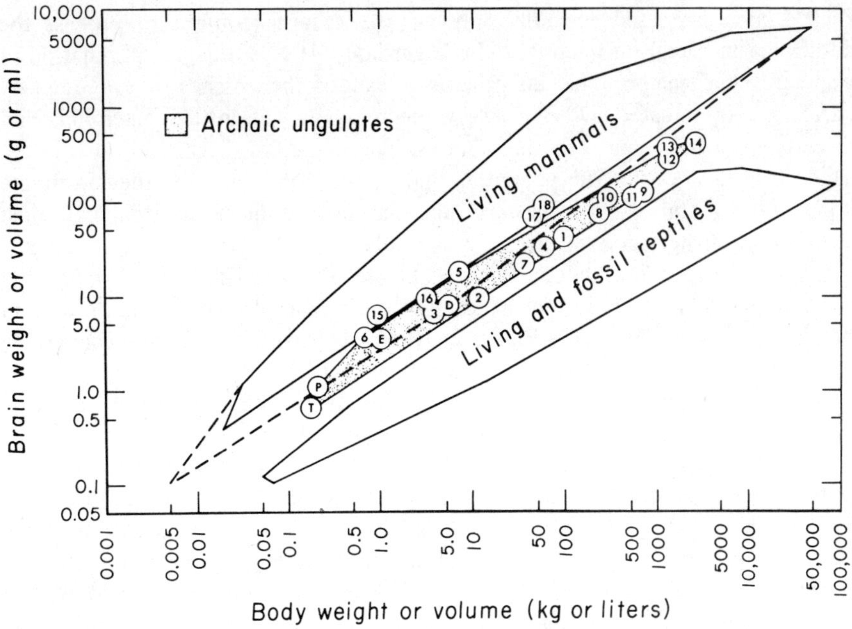

FIG. 1 Brain–body relations in the evolution of mammals from reptiles. Minimum convex polygon for living and fossil reptiles "maps" a reptilian grade of brain evolution achieved by immediate ancestors of mammals as well as by dinosaurs and flying reptiles. Archaic ungulates (stippled area, circled Nos. 1–14) and carnivores (order Creodonta, Nos. 15–18), and oldest known mammalian "brain" and body data for *Triconodon* of 150 million years ago (letter T) are in advance of the reptilian level and at the lower limit of or slightly below the living mammalian level. (From Jerison, 1973).

below the lower border of the polygon of living mammals. It is parallel to the reptilian and mammalian polygons and suggests a status achieved by the earliest mammals of 150 million years ago as well as by the first great radiation of mammals in the early Tertiary.

Without further analysis, Figure 1 indicates that before the relative brain size of living mammals evolved, a broad adaptive radiation of the archaic mammals at the beginning of the Tertiary took place with little or no gain in relative brain size beyond that achieved by the earliest Mesozoic mammals. A single polygon, similar in shape to that of living reptiles and mammals and oriented similarly in the coordinate plane, maps the level of brain development achieved in the early Tertiary by the archaic ungulates. That level is at or below the lower limits of the mammalian data of Crile and Quiring (1940), and when the creodonts are excluded there is no overlap at all between the polygon for archaic mammals and that for living mammals in the sample described by Crile and Quiring (solid line). The conclusion is, essentially, that in the early radiation of the mammals, the archaic forms invaded many niches by evolving larger and specialized bodies but their brains increased in bulk only to the extent required to process the additional information to control the larger body. It is as though a population of opossums and small-brained insectivores expanded their niches by evolving new species that increased in body size while remaining magnified didelphids and insectivores with respect to their brains. The degree of magnification is established by the slopes of the polygons, that is, their orientation in the coordinate plane, which states the necessary minimal change in brain size that must accompany changes in body size.

To provide a sense of the appearance and relative body sizes of some of these animals, lifelike reconstructions of some of the species of Figure 1 are drawn and numbered in the same way in Figure 2. These reconstructions have fascinated me for two reasons. First, they show pretty well that the archaic forms can be visualized as quite similar in body dimensions and skeletal adaptations to living mammals. Second, and as a corrolary to the first observation, the archaic forms can readily be visualized as living within niches comparable in many ways to those of living animals. These observations are not entirely unbiased, of course, because the artists and scientists involved in making the reconstructions were obviously influenced by the anatomy and appearance of living mammals. The skeletons of the archaic fossils have to be examined with care and with a practiced eye to verify the fact that they are, indeed, of primitive animals. The illustrated reconstructions were made quite carefully, with the reconstruction moving outward, in some instances, from muscular sheaths to efforts at indicating skin and sense organs appropriately (e.g., Osborn, 1929). The only serious errors would arise if a species had developed an unusual adaptation of soft tissue, such as the camel's hump, which is not clearly mirrored in the skeleton. The particular reconstructions in Figure 2 are all based on rather complete material and involve few guesses about the skeleton.

STRUCTURAL ENCEPHALIZATION: FOREBRAIN AND HINDBRAIN

In the illustrations of the endocasts of archaic Tertiary mammals such as *Phenacodus* in Figure 3 and *Arctocyon* in Figure 6, below (specimens 4 and 1, respectively, in Figure 1), there is evidence that the brains of most early mammals were unflexured. It is, therefore, possible to estimate the volumes of at least three subdivisions, forebrain, hindbrain, and olfactory bulbs. The method is illustrated in Figure 3. Using these estimates, it is possible to analyze the evolution of the relative sizes of these three parts of the brain by comparing them in the archaic species with those of later, progressive, species of mammals.

Although there are scattered data in the anatomical literature on the sizes of the parts of the brain in living mammals (e.g., Latimer, 1956; Stephan *et al.*, 1970) most of the literature is restricted to relatively few species and to relatively small animals, such as insectivores. A major study by Wirz (1950)

FIG. 2 Reconstructions of selected archaic ungulates to indicate their body configurations. Numbers as in Figures 1, 4, 5, and 7 refer to following species: (4) *Phenacodus primaevus;* (5) *Meniscotherium robustum*; (6) *Hyopsodus miticulus*; (7) *Pantolambda bathmodon*; (9) *Barylambda* sp.; (10) *Coryphodon hamatus*; (12 and 13) *Uintatherium anceps*. Oldest species (Middle Paleocene) at top and latest (Upper Eocene) at bottom. Note scale. (From Jerison, 1973.)

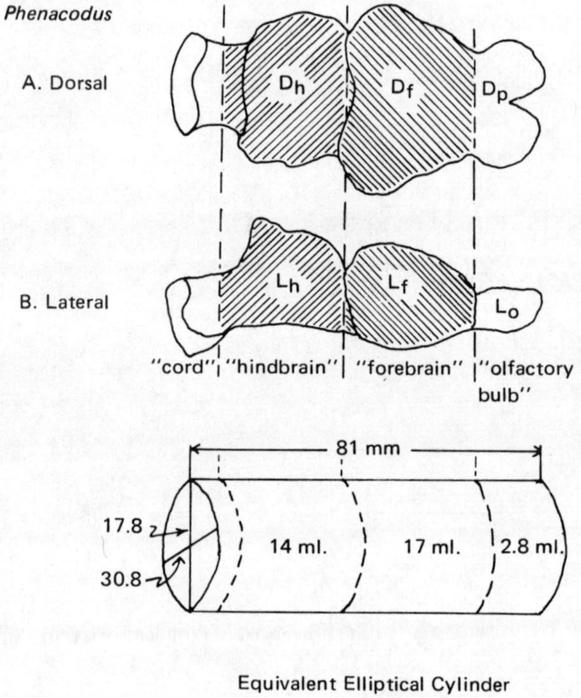

$$
\begin{aligned}
V_{total} &= abl\pi \text{ mm}^3 \\
&= \left(\tfrac{17.8}{2}\right)\left(\tfrac{30.8}{2}\right)(81)\,\pi \\
&\cong 35{,}000 \text{ mm}^3 \\
&= 35 \text{ ml.}
\end{aligned}
$$

FIG. 3 Method of estimating volume of endocast illustrated with a sketch of the fossil endocast of *Phenacodus primaevus*. (From Jerison, 1973.)

covered the broad range of mammals required for the analysis, and I used her equations to estimate the sizes of forebrain and hindbrain. I measured the olfactory bulbs directly from endocasts of Pleistocene fossils, which are similar to Recent species. (See Jerison, 1973, for a complete discussion.)

The forebrain sizes of the archaic Tertiary sample are compared with those of living mammals in Figure 4. As in the analysis of total brain size, these are shown as minimum convex polygons enclosing the data points in log–log brain:body space. It can be seen in Figure 4 that the forebrain:body map of the archaic group is a minimum convex polygon similar in orientation to, but below, that of living mammals (lowercase letters a–g). The forebrain of living mammals is about four to five times as large as that of the archaic ungulates, according to the samples that are compared here. The forebrains of creodonts (numbers 15–18),

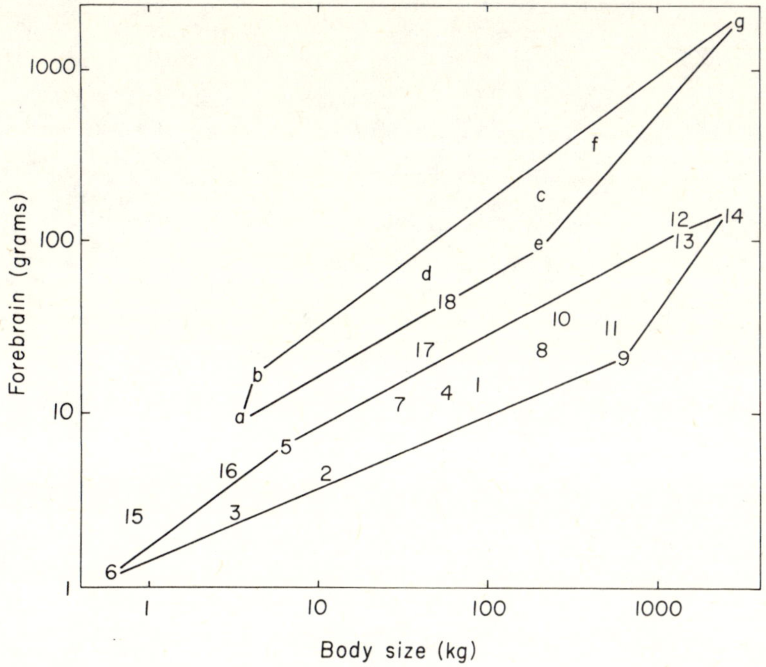

FIG. 4 Forebrain: body maps (minimum convex polygons) illustrating the evolution of the relative size of the forebrain from the archaic ungulate level below to that of living mammals above. From Jerison (1973).

like the brain as a whole, are about 50% larger than those of the archaic ungulates and overlap the forebrains of living species.

It is significant that these polygons, like those for the brain as a whole, maintain their orientation with a slope of about $2/3$ on the log–log coordinate system. This suggests that the body size factor, which may be thought of as the requirement that an extended body "surface" be mapped on the brain, holds true for the forebrain as well as for the brain as a whole. Particularly significant for the analysis of gross brain size as an approach to the general problem of brain evolution is the fact that the results with this somewhat finer analysis are consistent with the more gross earlier results based on the brain as a whole. From the study of the volume of the functionally "highest" regions of a fossil brain, essentially the same results are reached as those achieved earlier with the more gross data on the whole brain, which are normally all that are available for fossils.

The portions of the endocasts of archaic mammals described as hindbrain in Figure 3 are compared with data for living mammals and are presented in Figure 5. It is apparent, on inspection, that these data are almost as orderly as those of the previous graphs of brain:body relations. Except for the creodonts, the

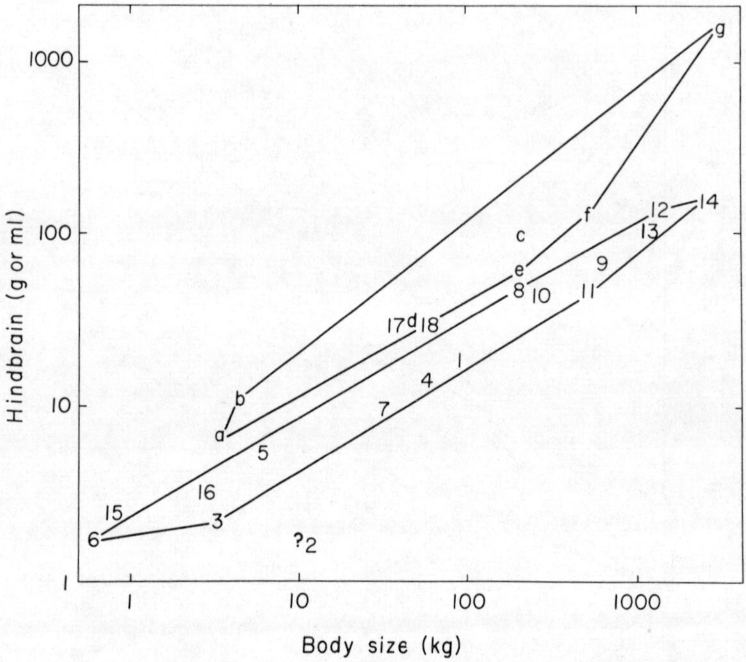

FIG. 5 Hindbrain: body map; cf. Figure 4. (From Jerison, 1973.)

archaic mammals have consistently smaller "hindbrains" than living mammals. The creodonts appear to be at the lower part of the range of the mammals from the living sample. [The hindbrain of the Upper Paleocene *Arctocyonides*, point No. 2, was probably reconstructed incorrectly in the reconstructions by Russell & Sigogneau (1965) which I used for its measurement.] The general orientation of the hindbrain polygons is also at a slope of about $^2/_3$. There was some increase in relative size of the hindbrain in creodonts and perhaps again between creodonts and living mammals. The increase was comparable to that for the forebrain and for the brain as a whole.

It should be emphasized that the result in Figure 5 is more surprising than the previous results. It indicates that the hindbrain (in particular cerebellum, which contributes most of the volume) increased in size in about the same way and proportion as did the forebrain. The evolution of brain size from the Eocene to the present was not merely "corticalization." The hindbrain was apparently involved to about the same extent as the forebrain.

STRUCTURAL ENCEPHALIZATION: THE OLFACTORY BULBS

In archaic endocasts, such as that of *Phenacodus primaevus* (Figure 3) or *Arctocyon primaevus* (Figure 6), the *relatively* large olfactory bulbs may suggest that "enlarged" olfactory areas are diagnostic of primitiveness in a brain. But

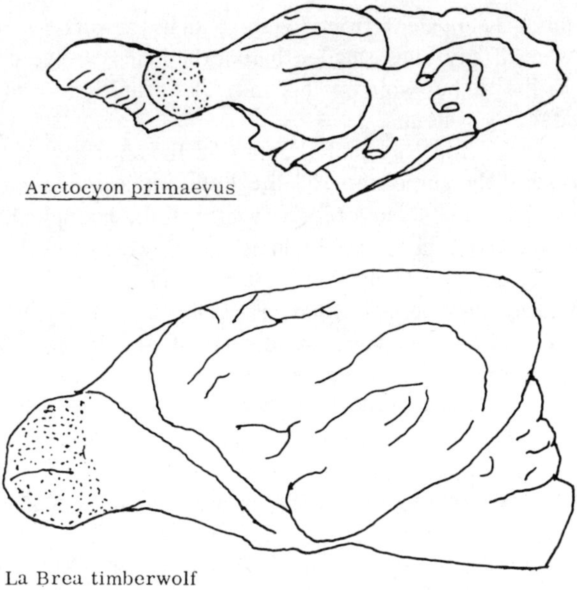

FIG. 6 Lateral views of endocasts of *Arctocyon primaevus* (Condylarthra, 60 million years ago) and the La Brea timberwolf, *Canis lupus* (Carnivora, about 50,000 years ago) to illustrate differences in absolute volume of the olfactory bulb regions shown as the stippled areas. (From Jerison, 1973.)

what does this really mean? A brain or endocast examined without regard to absolute dimensions is examined without regard for the biological significance of size. In the case of the olfactory bulbs, which have been believed to contain mainly peripheral "receivers" in the neural apparatus that processes olfactory information (Adey, 1959), a certain amount of information can be or must be received, and this should be independent of the size of the animal or of the rest of its brain. Olfaction is not analogous to sensory–motor coordination in which a larger animal with more muscle fibers must have more neural elements to control the fibers and must, necessarily, have a larger nervous system to contain those neural elements. In the case of the olfactory bulbs the important thing should be the absolute rather than the relative size. This comparison for the endocast of *A. primaevus* with that of the endocast of a living timber wolf (*Canis lupus*) is shown in Figure 6.

The olfactory bulb region of *Arctocyon* as preserved in the endocast (Russell & Sigogneau, 1965) includes some of the turbinates, and the region as a whole is somewhat enlarged in the endocast. However, the limits of the bulbs (stippled in Figure 6) are fairly clear, and their volumes can also be estimated by the method of graphic double integration illustrated in Figure 3. For *A. primaevus* the total volume of the bulbs was 4.1 ml. Whether this is a relatively small, normal, or large structure cannot be determined unless the expected volume is known. The timber wolf is a living carnivorous mammal of about the same body size as

Arctocyon, and it is apparent from Figure 6 that the olfactory bulbs in the archaic form were, if anything, smaller than in the living carnivore. (The volume of the bulbs in the timber wolf was 5.2 ml.) The primitive *Arctocyon* did not have a larger olfactory system.

The analysis for the olfactory bulbs cannot be the same as that for the rest of the brain, because the impression of the bulbs in an endocast is normally considerably enlarged. In the coyote, *Canis latrans*, for example, I measured the volume of the olfactory bulbs in the brain as 1.5 ml, whereas in the endocast the bulbs displaced 4.2 ml. It is likely that there is a consistent relationship between these two measures, however, and that mammals with truly large neural olfactory bulbs are also likely have large impressions of the olfactory bulbs on their endocasts. The comparison is therefore between the volumes of the olfactory bulbs taken from endocasts for the archaic Tertiary sample and the volumes in a special sample of Quaternary mammals selected for this purpose. The latter sample consists of fossils from the La Brea tar pits and a few living mammals.

Data on the size of the olfactory bulbs were related to body size and graphed in brain:body space. Minimum convex polygons could then be drawn about the data of the archaic Tertiary forms and of the progressive Quaternary species. This comparison is illustrated in Figure 7. (The letters a–g refer to different species from those shown in Figures 4 and 5.)

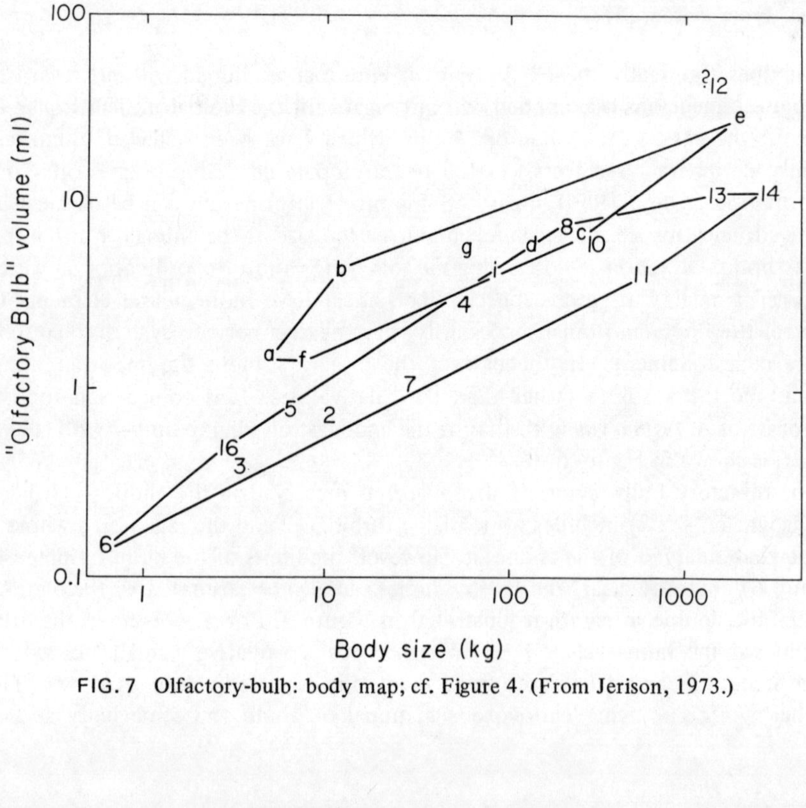

FIG. 7 Olfactory-bulb: body map; cf. Figure 4. (From Jerison, 1973.)

The results are unequivocal. The archaic mammals did not have relatively larger olfactory bulbs than recent mammals. If there is a difference related to the 50 million years that separates these groups, it is in favor of the later group as having relatively larger olfactory bulbs.

I have excluded one individual of the Early Tertiary specimens, *Uintatherium anceps* (No. 12), from the lower polygon; a second individual of that species (No. 13) is represented and has much smaller olfactory bulbs. Careful inspection of the endocast of Specimen 12 convinced me that the olfactory bulbs had been incorrectly reproduced (see Marsh, 1886, Plate VI), and that the anterior portion of the endocast was a result of an excavation of the cribiform area in which the ethmoidal bones had been removed. The anterior portion of the olfactory bulbs in that specimen was an artifact, an error in preparing the endocast.

The orientation of the olfactory-bulb:body polygons still indicated an exponent of about $2/3$ for the functional relationship. This means that the size of the olfactory bulbs is a function of body size and that body size must be taken into account in estimating the significance of the size of the olfactory bulbs. Sacher (1970) has indicated that this could have been anticipated, but that the contribution of a body size factor to the size of the olfactory bulbs may be a result of different causal variables than its contribution to the total brain size or the size of the forebrain and hindbrain.

UNIFORMITIES AND THE CONSERVATION PRINCIPLE

This report has been unusual in that orderly differences among species and among time periods have been discussed and methods for analyzing these differences have been emphasized. In most behavioral, physiological, and anatomical research on the brain, differences among species are ignored and the uniformities are emphasized. The researcher usually chooses to work with a species for reasons of convenience, and other species may be tested only to be sure that principles discovered in one species are not "species specific," that is, that they represent "general" principles of brain functioning. In fact, most principles of brain function are "general" principles rather than "species specific," despite the differentiation and diversification that are the rule of evolution. This paradox is worth examining more carefully, because it is resolved by an important evolutionary principle, the "conservation principle." Few neuroscientists and students of behavior have realized that this principle is the reason they can normally ignore problems of species-specific structures and functions.

The conservation principle was well stated by Stebbins as "the conservation of organization":

> Whenever a complex, organized structure or a complex integrated biosynthetic pathway has become an essential adaptive unit of a successful group of organisms, the essential features of this unit are conserved in all of the evolutionary descendants of the group concerned [Stebbins, 1969, pp. 124–125].

In this discussion this principle has been used in arguing the case that either the neuron or the "miniature nervous system" is a very old evolutionary achievement. It was also part of the basis for rejecting encephalization of function as a likely evolutionary phenomenon, if "function" is defined rigorously. The validity of this principle can be recognized in the analysis of the "wiring diagram" of the nervous system, because of the similarities in the basic plan of that diagram, e.g., among placental mammals, among mammals generally, and even among vertebrates, as this is made clear in discussions of homologies.

There are several ways in which the principle appears to be relevant for the analysis of the fossil record, including the analysis of relative brain size and speculations about the evolution of behavior. One example is the ubiquity of the exponent $^{2}/_{3}$ in brain:body maps (e.g., Figure 1). This exponent must result from some fundamental aspect of the organization of the nervous system in large and small vertebrates, and it is usually assumed to be related in some way to surface–volume relationships. The volume of the endocranial cavity or of the brain is proportional to the surface of the body, suggesting a mapping of the body on the brain of the type familiar in reports of localization of function. A more detailed statement of the source of the exponent should be possible, and it should be possible to develop a kind of statistical mechanics–a statistical morphology–that can derive gross brain:body relations from elementary relations between neuronal elements and the bodily structures and functions that they control.

The conservation principle suggests that such a statement should include parameters associated with taxonomic level. The statement might be written differently to describe a genus, family, order, or class, and experience with brain:body relations suggests that the statement with a $^{2}/_{3}$ exponent applies only at taxonomic levels of a class or order (cf. Figure 1). At lower levels, such as the genus or even the family, the exponent is of the order of $^{1}/_{3}$ or less (Jerison, 1970a); within a species of mammals the exponent may even be zero (Sholl, 1948). There is no really adequate explanation for this changing pattern, and it is probably impossible to make an adequate analysis without detailed quantitative data on the relations among structures of the brain for many different species and for individuals of different body size "within-species."

I have used the conservation principle in various ways in theorizing about the evolution of enlarged brains (Jerison, 1973). One of these arguments may be instructive in that it shows how a very simple application of this principle and the principle of proper mass can lead to some insight into the evolution of neural control systems. I have argued that most of the identified functions of the brain that are at all well understood are functions of very minute amounts of tissue, and current knowledge of the brain in the large is surprisingly limited and paradoxical. Extensive brain damage can have apparently trivial effects on measured behavior (Hebb, 1949), for example, while the activation of very small regions of the brain can be surprisingly effective in eliciting elaborately defined behavior patterns (Valenstein, Cox, & Kakalewski, 1970). When it is asserted that evolutionary changes in the size of the brain have been associated with certain behavioral or physiological adaptations, there should be some evidence that large masses of nervous tissue have actually been required as parts of the mechanisms of those adaptations. The argument is, essentially, that truly new organizations (in Stebbins' sense) must be demonstrated for which the "proper mass" of neural tissue is large.

As a concrete illustration, consider the evolution of endothermy (see also Nyberg, 1971; Jerison, 1971a). Adaptations for the intrinsic control of body temperature are certainly among the more dramatic of the differentiating features of the vertebrates. The living lower vertebrates, as defined by brain: body maps, are almost all ectotherms, basing their control of body temperature on extrinsic mechanisms. The endotherms, that is, the birds and mammals, are considered more advanced in their control of body temperature because of the intrinsic mechanisms that they have evolved. In fact, there are many grades between purely intrinsic and purely extrinsic controls (Heath, 1968), and it is quite clear that reptiles as well as birds and mammals are generally "warm blooded" when they are active. They differ in the way they control body temperature. From the viewpoint of the role of the brain in these processes, animals with extrinsic temperature control, such as lizards, which make major behavioral responses by moving into or out of sunny and shady areas when they sense appropriate changes in body temperature, may use more neural tissue than endotherms in handling the control of body temperature. The evidence on brain mechanisms for endothermy indicates that only a small amount of neural tissue in the hypothalamus is actually involved in the control (see, e.g., Myers & Veale, 1970). Despite these facts, ectotherms are not larger brained than endotherms. The conclusion that endothermy has been a cause of brain enlargement cannot be accepted *a priori*. A more appropriate conclusion is that the control of body temperature, as a dimension of adaptive change, was irrelevant for the evolution of encephalization.

I have used the conservation principle in more positive ways to speculate on factors affecting changes in brain size, by assuming that a method of processing sensory information in one modality (e.g., vision) was copied in important ways in the evolution of another modality (e.g., audition). I shall not repeat the argument here (see Jerison, 1971a, 1973). Although developed from considerations of adaptations for processing distance information that had to occur in the evolution of the earliest mammals as nocturnal "reptiles," the analysis is also consistent with recent experiments on the organization of receptor fields in the auditory, visual, and tactile areas of the cerebral cortex. The conservation principle is consistent with the evidence of the evolution of brain size. This evolution was conservative in that encephalization was uncommon in vertebrate evolution. Of the approximately 40,000 living species of vertebrates only the mammals and birds, and a few sharks (and perhaps some electric fish)–about 15,000 species–are encephalized beyond the most primitive grade (Jerison, 1973).

SELECTION PRESSURES

Among the necessary activities of evolutionists are their analyses of selection pressures that are effective in directing evolutionary change. Properly understood, such selection pressures are the basis of the apparently orderly and directed diversification or progress in the organization of brain and behavior in the phyletic history of living organisms. In the case of the evolution of behavior a student is really limited mainly to such an analysis in any effort at reconstructing the actual history of the evolution of behavior mechanisms. The analysis may be tested by correlating it with the fossil record of endocranial evolution and of the evolution of skeletal structures associated with the particular behaviors.

There has not been very much systematic analysis of this type. I have attempted it in my analysis of the requirements for sensory control, imagery, and intelligence as vertebrate species entered various adaptive zones, and I treat these as selection pressures associated with the enlargement of the brain (Jerison, 1973, 1975). Behavioral studies that are cast in an evolutionary or comparative framework can rarely avoid some circularity in reasoning. If a particular "fixed action pattern" is observed, it must be adaptive to some niche, and the mechanisms for it must have evolved in association with the selection pressures of the niche. But the judgment that the behavior is adaptive is usually based on the fact that it was observed, not on the analysis of the niche. The analysis, if made, follows the observation. It is not my intention to point ways out of this dilemma, only to mention it in order to emphasize some of the difficulties inherent in this important aspect of evolutionary analysis.

The analysis of selection pressures that resulted in the evolution of neural control and of the nervous system is less common than the comparable analysis of behavior. My own effort and those of others (e.g., Polyak 1957) require a difficult compromise between armchair natural history and analysis that can lead to useful insights into the history and present status of the nervous system. One problem is how to phrase the crucial questions and this can be resolved with the help of quantitative methods. To illustrate, the data of Figure 1 suggest the question: how universal was the apparent trend among mammals toward increased brain size from the level of archaic mammals of the Mesozoic and the early Tertiary to the present? The answer to this question might not identify selection pressures, but it would suggest whether these pressures crossed the boundaries of the various adaptive zones that distinguished the orders of mammals.

A rather clear and (for me) surprising answer to that question is possible if encephalization (measured by the encephalization quotient,[1] *EQ*) in living mammalian orders is considered. The cumulative frequency distributions of *EQ* in insectivores and rodents are compared with that of the archaic ungulates in Figure 8. It is obvious that there is a difference even between the "primitive" living insectivores and the archaic condition. There is even more encephalization evident in the data on species of living rodents. Selection pressures toward increased encephalization were evidently widespread in mammals. But Figure 8 also shows that niches for relatively small-brained species exist for both insectivore and rodent species, because of the overlapping of the distributions at their lower ends. The evidence is that the selection pressures led to increased diversification in relative brain size. There appears to be a lower limit to mammalian brain size at about $EQ = 0.1$ (at one-tenth of the expected brain size for a given body size); therefore, diversification can only be manifested as increased brain size above that lower limit.

Perhaps the most elegant analysis of the role of selection pressures in the evolution of the brain was that by Welker and his colleagues (Welker & Seidenstein, 1959; Welker & Campos, 1963), who examined localization of function in the raccoons and their relatives. They were able to show a clear correlation of the amount of tissue in the motor cortex and the degree of its differentiation in five genera of procyonids with the extent to which the forepaws were used as hands to manipulate objects. This result also illustrates the principle of "proper mass."

The importance of selection pressures in determining the direction of evolution of the brain can be acknowledged by stating explicitly a final principle (Jerison, 1973, p. 15):

> The brains of all species evolved in ways appropriate to control the behavior required for life in their environmental niches.

This principle is, essentially, a statement that the brain evolved under natural selection. It is the basic principle of brain evolution and may be termed the "principle of adaptiveness." To apply it one must charactize an environmental niche completely. This fundamental task of evolutionary biology is also funda-

[1]The encephalization quotient *EQ* is discussed in Jerison (1970b, 1973). The exact definition of EQ_i for species i is

$$EQ_i = E_i/0.12P_i^{2/3}$$

where E_i is a measured (or typical) brain weight or volume in grams or milliliters and P_i is the corresponding body weight or volume in the same units. The denominator of the equation is the formula for "expected" brain size in "average" living mammals; that formula is a "best fitting line" through the upper polygon of Figure 1. The encephalization quotient *EQ* is a dimensionless constant giving a quick estimate of the degree of structural encephalization in a particular species, more or less independently of body size.

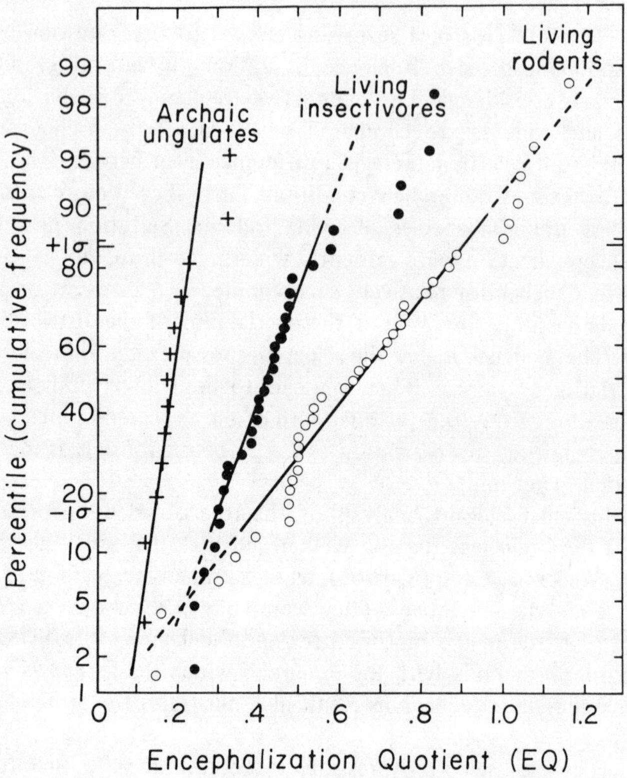

FIG. 8 Cumulative frequency distributions of encephalization quotients *EQ* in mammalian orders representing a primitive fossil "Mesozoic" status (archaic ungulates) and living orders (Insectivora and Rodentia) occupying ecological niches that are assumed to be similar to those of Mesozoic mammals. *EQ* is defined as the measured brain size divided by the expected brain size for an average living mammal of a particular body size. (From Jerison, 1973.)

mental for the analysis of the evolution of the brain, which requires, in addition the description of the brain:behavior systems necessary for adaptations to the niche.

SUMMARY AND CONCLUSIONS

A number of evolutionary principles are relevant for the study of brain and behavior. Of the "standard" principles cited in textbooks, only the principle of encephalization with regard to the structure of the brain is clearly relevant, and the concept of encephalization of function is rejected. It is replaced by the concept of "conservation of organization," which asserts that functional adaptations tend to be conserved once they have evolved. Additional principles developed in this chapter are the principle of "proper mass" and the principle of "adaptiveness" of brain evolution. These refer (1) to the relative amount of brain present in a species as appropriate to the relative importance of the behaviors in the life of the species, and (2) to the general idea that the brain evolved under natural selection as part of the adaptations of species of animals to their environmental niches. These principles may be contrasted with conventional views of brain evolution in which living animals are arranged according to a "scala naturae" as representatives of some grade of evolution of the brain. The approach adopted in this chapter emphasizes the natural niche and habitat of a species as determining the evolution of its brain.

ACKNOWLEDGMENTS

This research has been supported by the State of California Department of Mental Hygiene and U.S. Public Health Service Grant No. HD-04612. Much of the text and all of the illustrations in this chapter are from Jerison (1973) and are reproduced with the permission of the publisher.

REFERENCES

Ades, H. W. Central auditory mechanisms. In H. W. Magoun (Ed.), *Handbook of physiology: Section I, Neurophysiology*, Vol. 1. Washington, D.C.: American Physiology Society, 1959, Pp. 585–613.

Adey, W. R. The sense of smell. In H. W. Magoun (Ed.), *Handbook of physiology: Section I, Neurophysiology,* Vol. 1. Washington, D.C.: American Physiological Society, 1959. Pp. 535–548.

Bauchot, T., & Stephan, H. Donées nouvelles sur l'encéphalisation des insectivores et des prosimiens. *Mammalia*, 1966, **30**, 160–196.

Bok, S. T. *Histonomy of the cerebral cortex.* New York: Van Nostrand, 1959.

Brazier, M. A. B. The historical development of neurophysiology. In H. W. Magoun (Ed.), *Handbook of physiology: Section I, Neurophysiology*, Vol. 1. Washington, D.C.: American Physiological Society, 1959. Pp. 1–58.

Bullock, T. H. Simple systems for the study of learning mechanisms. In F. O. Schmitt *et al.* (Eds.), *Neurosciences research summaries,* Vol. 2. Cambridge, Mass.: MIT Press, 1967. Pp. 203-327.

Crile, G., & Quiring, D. P. A record of the body weight and certain organ and gland weights of 3690 animals. *Ohio Journal of Science*, 1940, **40**, 219–259.

Grinnell, A. D. Comparative auditory neurophysiology of neotropical bats employing different echolacation signals. *Zeitschrift für Vergleichende Physiologie*, 1970, **68**, 117–153.

Harman, P. J. *Paleoneurologic, neoneurologic, and ontogenetic aspects of brain phylogeny.* James Arthur lecture on the evolution of the human brain. New York: American Museum of Natural History, 1957.

Heath, J. E. The origins of thermoregulation. In E. T. Drake (Ed.), *Evolution and environment,* New Haven: Yale University Press, 1968. Pp. 259–278.

Hebb, D. O. *The organization of behavior: A neuropsychological theory*. New York: Wiley, 1949.

Jerison, H. J. Brain evolution and dinosaur brains. *American Naturalist*, 1969, **103**, 575–588.

Jerison, H. J. Gross brain indices and the analysis of fossil endocasts. In C. R. Noback and W. Montagna (Eds.), *The primate brain, Advances in primatology,* Vol. 1. New York: Appleton-Century-Crofts, 1970. Pp. 225–244. (a)

Jerison, H. J. Brain evolution: new light on old principles. *Science*, 1970, **170**, 1224–1225. (b)

Jerison, H. J. More on why birds and mammals have big brains. *American Naturalist*, 1971, **105**, 185–189. (a)

Jerison, H. J. Quantitative analysis of the evolution of the camelid brain. *American Naturalist*, 1971, **105**, 227–239. (b)

Jerison, H. J. *Evolution of the brain and intelligence*. New York: Academic Press, 1973.

Jerison, H. J. Fossil evidence of the evolution of the human brain. *Annual Review of Anthropology* 1975, **4**, 27–58.

John, E. R. Switchboard versus statistical theories of learning and memory. *Science*, 1972, **177**, 850–863.

Kandel, E. R. Cellular studies of learning. In G. C. Quarten, *et al*. (Eds.), *The neurosciences.* New York: Rockefeller University Press, 1967. Pp. 666–689.

Kennedy, D. Small systems of nerve cells. *Scientific American*, 1967, **216** (May), 44–52.

Lartet, E. De quelques cas de progression organique verifiables dans la succession des temps géologiques sur des mammifères de même famille et de même genre. *Comptes Rendus Acad. Sci.*, 1868, **66,** 1119–1122.

Lashley, D. S. *Brain mechanisms and intelligence.* Chicago: University of Chicago Press, 1929.

Lashley, K. S. Persistent problems in the evolution of mind. *Quarterly Review of Biology* 1949, **24,** 28–42.

Lashley, K. S. In search of the engram. In *Symposia of the Society for Experimental Biology* No. 4. Cambridge: Cambridge University Press, 1950. Pp 454–482.

Latimer, H. B. The weights of the parts of the brain in several species of animals. *Transactions of the Kansas Academy of Science*, 1956, **59,** 432–441.

Marquis, D. G. Effects of removal of the visual cortex in mammals, with observations on the retention of light discrimination in dogs. *Ass. Res. Nerv. Ment. Dis.*, 1934, **13**, 558–592.

Marsh, O. C. Small size of the brain in Tertiary mammals. *Amer. J. of Sci. Arts,* 1874, **8,** 66–67.

Marsh, O. C. *Dinocereta*. Monographs of the United States Geological Survey, No. 10, 1886, pp. 1–243.

Myers, R. D., & Veale, W. L. Body temperature–possible ionic mechanism in the hypothalamus controlling the set point. *Science*, 1970, **170**, 95–97.

Noback, C. R., & Moskowitz, N. Structural and functional correlates of "encephalization" in the primate brain. *Annals of the New York Academy of Science*, 1962, **102,** 210–218.

Nyberg, D. An hypothesis concerning the larger brains of homoiotherms. *American Naturalist*, 1971, **105**, 183–185.

Osborn, H. F. *The Titanotheres of ancient Wyoming, Dakota and Nebraska.* 2 Vols., United States Geological Survey Monographs, No. 55, 1929.

Polyak, S. (H. Kluver Ed.), *The vertebrate visual system*. Chicago: University of Chicago Press, 1957.

Rovainen, C. M. Physiological and anatomical studies on large neurons of central nervous system of the sea lamprey (*Petromyzon marinus*). I. Müller and Mauthner cells. *Journal of Neurophysiology*, 1967, **30**, 1000–1033.

Ruch, T. C. Cortical localization of somatic sensibility. The effect of precentral, postcentral and posterior parietal lesions upon the performance of monkeys trained to discriminate weights. *Research Publications of the Association for Research in Nervous and Mental Diseases*, 1935, **15**, 289–330.

Russell, D. E., & Sigogneau, D. Étude de moulages endocrâniens de mammifères Paléocènes. *Mémoires, Muséum National d'Histoire Naturelle* (*Paris*), **14** (n.s. serie C). 1965. Pp. 1–36.

Sacher, G. A. Allometric and factorial analysis of brain structure in insectivores and primates. In C. R. Noback & W. Montagna (Eds.), *The primate brain*, (*Advances in primatology*, Vol. 1. New York: Appleton-Century-Crofts, 1970. Pp. 245–287.

Sholl, D. A. The quantitative investigation of the vertebrate brain and the application of allometric formulae to its study. *Proceedings of the Royal Society of London. Series B*, 1948, **135**, 243–258.

Stebbins, G. L. *The basis of progressive evolution*. Chapel Hill, N.C.: University of North Carolina Press, 1969.

Stephan, H., Bauchot, R., & Andy, O. J. Data on size of the brain and of various parts in insectivores and primates. In C. R. Noback & W. Montagna (Eds.), *The primate brain, Advances in primatology,* Vol. 1. New York, Appleton-Century-Crofts, 1970. Pp. 289–297.

Tilney, F. Fossil brains of some early Tertiary mammals of North America. *Bulletin of the Neurological Institute, New York,* 1931, **1**, 430–505.

Valenstein, E. S., Cox, V. C., & Kakalewski, J. W. Reexamination of the role of the hypothalamus in motivation. *Psychological Reviews,* 1970, **77**, 16–31.

Weiskrantz, L. Encephalisation and the scotoma. In W. H. Thorpe & O. L. Zangwill (Eds.), *Current problems in animal behavior.* London: Cambridge University Press, 1961. Pp. 30–58.

Welker, W. I., & Campos, G. B. Physiological significance of sulci in somatic sensory cerebral cortex in mammals of the family Procyondiae. *Journal of Comparative Neurology*, 1963, **120**, 19–36.

Welker, W. I., & Seidenstein, S. Somatic sensory representation in the cerebral cortex of the raccoon (*Procyon lotor*), *Journal of Comparative Neurology,* 1959, **111**, 469–502.

Wever, E. G. Structure and function of the lizard ear. *Journal of Auditory Research*, 1965, **5**, 331–371.

Wilson, D. M. The flight-control system of the locust. *Scientific American*, 1968, **218**(5), 83–90.

Wirz, K. Zur quantitativen Bestimmung der Rangordung bei Säugetieren. *Acta Anatomica*, 1950, **9**, 134–196.

Woolsey, C. N. Organization of somatic sensory and motor areas of the cerebral cortex. In H. F. Harlow & C. N. Woolsey (Eds.), *Biological and biochemical bases of behavior.* Madison: University of Wisconsin Press, 1958. Pp. 63–81.

4

Rates of Evolution of the Nervous System and Behavior

Everett C. Olson

The University of California, Los Angeles

INTRODUCTION

The problems of rates of evolution of the nervous system and behavior remain refractory. They involve integration of three kinds of information–morphological, behavioral, and temporal. The time must be sufficient that evolutionary change can be followed. This means that much of the primary data must come from fossils. Functional interpretations must therefore be based on analogies with living animals and long and tenuous lines of inference are necessary. Results pertaining to behavior depend on the often uncertain links between structure and behavior in living organisms and their transference to fossils. They are inevitably less than satisfying. Testability on primary materials remains at a minimum. This chapter is designed to present a balanced analysis of the problems; what has been accomplished; what can be accomplished, at least with present methods and concepts; and what cannot be accomplished.

The format involves a summary of the results of work on rates of evolution and on behavior based primarily on the fossil record. This is followed by an overview of the problems in light of availability of information on morphology of the nervous system and rates. Then the current status of the subject is taken up using summaries of some recent studies and interpretations as examples to show both the strengths and weaknesses of current knowledge.

It becomes apparent that persistence of the problems comes in part from an inability to ask pertinent questions that have some hope of testable answers. Although this is partly a matter of limitations of the materials, it appears to stem as well from the immense difficulties that any one student faces in bridging the gulf between necessarily reductionist analyses, the holistic nature of many behavioral problems, and the special nature of the temporally disturbed but limited information available from fossil materials. Throughout this chapter, a special effort is made to clarify the last point and to show where syntheses with other approaches may be fruitfully consummated.

THE MATTER OF RATES

The subject of rates in organic evolution was treated thoroughly and authoritatively by Simpson in 1944 and 1953. His 1944 study was followed by a strong upsurge of interest in this field coordinated with the rising interest in a synthesis of genetics, systematics, and paleontology. Stimulated by various students' studies and by Simpson's later work in 1953, the subject of rates of evolution remained vigorous through the first half of the 1950s. Particularly active during this period and somewhat later was Kurtén (see, for example, Kurtén, 1959). Sporadic studies during the 1960s dealt specifically with problems of rates; however, for the most part the subject played a persistent but secondary role as one aspect of phylogenetically oriented paleontological analysis.

Relatively little has been added to the theoretical aspects of the study of rates of evolution outlined in Simpson's studies. The scope of materials treated has expanded. Some different ways of assessing rates have been proposed, and in special fields closer integration between the data from fossils and that from living organisms has been attained. Knowledge has been increased but with little change in the basic aspects of the field of study.

Simpson's works showed three basic needs for success in any study of rates:

1. establishment of a time scale;
2. definition of the entity being studied, e.g., a character, a suite of characters, a system, relative change of two or more characters (allometry), or a taxonomic category;
3. establishment of a suitable phylogeny under which the object of study defined under (2) can be assessed.

The establishment of a time scale involves two approaches. One is the use of the geological time scale, with unequal time units; the other is the use of a clocklike time scale, with equal time units. Since 1953 great strides have been made in both the extent and the precision of radioactive dating, which supplies a clocklike scale, and a strong shift to use of this temporal framework has ensued.

Evolutionary change can be viewed in many perspectives–genetic, biochemical, physiological, ontogenetic, morphological, behavioral, and so forth. Because evolutionary changes are ultimately genetic, efforts are continually directed toward relating change and rates of change to their genetic background. Because the first-order evidence in the fossil record is morphological, Simpson (1944, 1953), Haldane (1949), and Kurtén (1959) made special efforts to relate morphological and genetic change in phyletic sequences. This is, of course, implicit in the work of many others but only rarely is it explicit.

To approach this end, changes in a single character or a very limited suite of characters are generally used. In lieu of this, overall modification is judged by the use of some such taxonomic category as species or genus (most common) or some higher level. The ability to draw genetic inferences lessens with the increased level of the taxonomic category. Because of this limitation, the paleontological evolutionist often finds her- or himself in the position of conceiving his or her materials under population concepts but of dealing with them in an operationally typological framework. This has posed serious problems in actual studies of rates and the general relationships to evolutionary processes under neo-Darwinian theory.

Phylogenies, or lines of descent, are essentially best estimates of relationships based on currently available materials. They are subject to change and such changes can vitiate conclusions on rates, even those from which broad generalizations have developed. If phylogenies are very broadly framed, say in sequences of families or more inclusive categories, they may be expected to have considerable stability and to serve for broad analyses of directions and rates of changes. Genetic inferences, however, are dangerous. At specific or generic levels, available for the most part only for Tertiary mammals, phylogenies provide framework in which morphological and functional changes can be more meaningfully related to evolutionary theory.

At this level, however, analyses have tended to shift to studies of rates of evolution of the categories themselves, the number of genera or of species that originate per given unit of time, for example. Although there are merits to this procedure and rates are readily and meaningfully determined in many cases, such analyses do not pertain to the subject of this discussion, for organ systems as such are not considered.

THE MATTER OF BEHAVIOR

In 1959, Romer published a comprehensive paper on the evolution of behavior as part of a volume *Behavior and Evolution* based on earlier conferences on the subject held in 1955 and 1956. He outlined thoroughly the gross aspects of evolution of behavior, emphasizing the major behavioral advances. Special attention was given to chordates, and to the vertebrates in particular, with less attention to the nonchordate invertebrates. The basis of his study, as far as possible, was the fossil record, but necessarily Romer used modern substitutes in his phylogenies when required and modern analogs throughout. He stressed the difficulties and fallacies of using a "phylogenetic" sequence based on living animals that included evolutionary "relicts." As others have done, he emphasized the great temporal and evolutionary distances of such organisms from their remote, if somewhat similar, ancestors.

This problem, of course, is basic to all phylogenetic studies of behavior based, as they must largely be, on living animals. The matter was treated specifically with regard to brains and behavior by Jerison (1969a). It arises constantly in phylogenetic analyses involving data of molecular biology as a basis for understanding rates of change from materials but little removed from their genetic base. In this area (e.g., Ohno, 1970) evidence of a much greater stability than may have been anticipated has emerged and can, of course, be carried over as an indication of great genetic stability into all studies of rates in which genetic heritability is the basic factor. Implications with respect to behavior certainly exist if behavior, as seems true at least in the sense of capacities to behave, is genetically determined. For the present, this sort of information lies far beyond the scope of any tangible analysis at the level of rates with which I am concerned.

Romer's study dealt implicitly with rates of change but the matter was treated only in a very general sense. Basic to his analysis was the proposition that evolution in general had moved toward increased complexity. General morphology and behavior kept stride in the course of this increase. For the most part, increase in complexity was gradual; however, when marked change in habitat occurred, the evolution of both morphology and behavior was accelerated.

Increasing complexity in behavior was related by Romer primarily to the development of locomotion and concomitant development of the sensory and nervous systems. The two aspects of function most available to study in the fossil record are feeding and locomotion. It is only natural that vertebrate paleontologists tend to view much of evolution in the light of these most tangible properties. This is in part, but only in part, theoretically justified, for keys to change, the initial shifts, may lie in other systems of the body or in behavior itself. Locomotion and feeding may follow rather than lead, adapting to other changes that have already occurred. Romer, of course, recognized this but he also recognized that knowledge of other systems among fossils was uncertain and, with rare exceptions, highly inferential.

Since the publication of Romer's analysis many other studies, usually more limited and specific, have appeared, just as many preceded his synthesis. Whereas Romer saw most changes as slow and behavioral changes as gradual, occurring over the course of many millions of years, other students viewed these sorts of shifts as occurring with relative rapidity. Earlier this idea was embodied in Simpson's concept of "quantum" evolution which, although changed in content and explanation over the years, expressed a concept of very rapid evolutionary shift. The shift among tetrapods from land to water, from lowland to upland, and from the reptilian habitus to the mammalian have all been suggested to have been rapid with the correlate of rapid behavior change. The last item is further explored in the final section of this chapter.

A second area of controversy that has important behavioral implications relates to Romer's concept of the origin of vertebrates in fresh water. He drew extensive inferences on form and function from this mode of origin. Others, especially Denison (1965), have argued to the contrary. If the opposite point of view is accepted, considerable modification of some aspects of Romer's model is required. It is this sort of problem with which those who have dealt with behavioral changes in vertebrate phylogeny have been most concerned. The behavioral aspects have been interpreted as forming the background of the study instead of themselves being its central thrust. There are major exceptions, particularly among some who have worked with the nervous systems, for example, Jerison (1963, 1969a, b, 1970a, b, 1971a), Edinger (1948, 1964a, b, 1966), Radinsky (1967), and many who have worked on primate and human paleontology, including Simons (1964) and Pilbeam (1972).

THE PROBLEMS OF RELATING THE NERVOUS SYSTEM, BEHAVIOR, AND THEIR RATES OF CHANGE

The relationship between structure and physiology of the nervous system and behavior can be treated definitively only by studies of living animals. What is learned can be applied to fossils to give a time scale for analysis of rate of change. This is effective to the extent that structures can be determined directly or inferred for fossils and that these fossils can be arrayed in phylogenies over significant periods of time. Each case that it may be desirable to study poses its own special set of problems. Some generalization on the availibility of various parts of the nervous system can, however, be made. They may be briefly summarized as follows.

Central Nervous System

The external shape of the brain and its volume are moderately to well reflected in endocranial casts of mammals. Proportions of these areas that have superficial expression can be determined, as can major sulci. Many students, foremost among them Tilly Edinger, have taken advantage of this fact to establish paleoneurology as a workable field. Among mammals, phylogenetic sequences of reasonable reliability can be erected and moderately accurate radioactive dating at close intervals is possible. This is easily the most fruitful area for research in the study of rates of evolution of the nervous system and behavior.

The brain is also well reflected in endocranial casts of birds. The fossil record of birds, however, is notably poor. Except for interpretation of conditions in the Jurassic, in *Archaeopteryx* (see Jerison, 1968, 1971b), some highly questionable information about Cretaceous birds, and comparisons of modern birds and reptiles, essentially nothing can be done to take advantage of the close relationship of nervous tissue and bone.

Among the other vertebrates, of course, the brain does not fill the cranial cavity so that only relatively crude interpretations can be made. Generally these deal with size, involving cranial capacity and the assumption that brain size and cranial capacity maintain a fairly constant ratio. This assumption is borne out to some extent by studies of modern lower tetrapods and fishes. There is room for considerable caution for reptiles near the reptilian–mammalian boundary as discussed in more detail on pages 54–63.

With these kinds of data, even accepting the presumed volumetric relations, only the grossest interpretations of the relationship of brain evolution and behavior can be made. It appears, as Jerison (1969a) has noted, that selective forces have had little effect on relative brain size prior to the changes that has led to birds and to mammals. However, specialized brain structures, related to specific habitats and activities of their bearers, have certainly appeared continually throughout the evolution of the nonmammalian and nonavian groups. For the most part these are uninterpretable in the fossil record.

The remainder of the central nervous system, the spinal cord, is surrounded by considerable nonneural tissue in the spinal canal of the vertebrae. Occasional expansions, such as that in the sacral region of some dinosaurs, are of passing interest with respect to interpretation of habits. Possibly, in such a case, if phylogenetic origins and changes could be traced, some points of interest relative to rates of changes of habits and morphological modifications might be found.

Other Portions of the Nervous System

The remainder of the nervous system is not generally available for study on the basis of direct evidence from fossil vertebrates. There are some exceptions. The exits of the cranial nerves are generally well shown in fossil skulls. The great stability of the number and positions of these nerves, however, reduces their utility for studies of behavioral changes. Size, number of foramina for an individual nerve, and position are the main ways in which they vary. Size has occasionally been used to interpret function; for example, the large outlet of the acoustic Nerve VIII in early whales (Edinger, 1955a). Mostly, however, the cranial nerve foramina have been used as stable landmarks for understanding homologies of associated cranial bones. It is largely among mammals, plus some fish and early amphibians, that the patterns of cranial nerve opening can be seen in the detail necessary for analytical studies. Some studies, for example those by Olson (1940) and Edinger and Kitts (1954) have dealt with them. Anything more than marginal utility for analysis of rates of behavior changes seems out of the question.

Various foramina of other nerves of the peripheral system occur in the skull and postcranium. It is primarily by an assumption of stability of the nervous system that the nerves passing through these foramina can be identified. From this fact alone it is evident that little can be done with them. Again in specific cases this may not be true, but these are sufficiently exceptional that the statement to the effect that utility of this part of the nervous system is extremely low holds true.

Sensory System

The various organs of special sense do leave well-defined records upon the bones. Some give access to interpretations of behavior and among these opportunities for establishing rates of change exist, although they are relatively few.

The lateral line, sensory system. Many fishes and some amphibians have clear records of the lateral line system on their skulls and scales. The nervous structures of the system receive low-frequency "near-field" components of water motion. They are good indicators of behavioral features of various fishes and amphibians and can be traced through time in well-established phylogenies. Behavioral interpretations of very early vertebrates have been made from them (e.g., Denison, 1966, and earlier). Little or nothing has been done with rates of change. This is certainly possible for later groups of fishes, among the actinopterygians, where adequate materials are potentially available. How significant results of such studies may be is open to serious question.

The sense of sight. The sizes, positions, and orientations of orbits provide some information on gross aspects of the sense of sight and on the behavior that has presumably aacompanied the inferred functions. In aquatic animals orbital positions contribute to the interpretations of the principal habitats, such as surface swimming or bottom living. Modifications of habits in adaptation to the various conditions can be traced through phyletic sequences in various instances. Little of this has been done and the amount of information added to what is more easily determined from other criteria is likely to be slight.

The size of orbits on occasions, as in ichthyosaurs, may be instructive and provide a guide for rates of change of behavioral patterns. The presence or absence of sclerotic plates suggests to some extent the stresses to which the eyeball is subjected. For the most part, however, the eye form of vertebrates being extremely stable, the fossil record does not record the kinds of more subtle information necessary for understanding the nature of vision itself. This must be done with living animals.

The pineal organ. The range of development of the pineal apparatus in extinct vertebrates, as indicated by the parietal fossa, offers considerable room for speculation on its relationship to behavior and changes of behavior in the course of adaptations to new circumstances. Considerable study has been made about the functions of the pineal structures (Steyn, 1961; Kelly, 1962, a summary; Stebbins & Eakin, 1958). Edinger (1956) treated the matter with attention to fossils. As a structure sensitive to external energy—light and heat—and one that is readily identified and followed through phyletic series in fossils, the pineal apparatus holds promise for interesting studies of concomitant changes of this part of the nervous system and behavior along with estimates of rates of change. The fossil evidence has not been marshalled as yet. Further comments on this matter are deferred until the final section of this chapter, where they can be put in evolutionary context.

The sense of smell. Estimates of the extent of the nasal epithelium and development of the olfactory bulbs of the brain can give insights into the sense of smell. Data for fossils are slight except for mammals and their immediate predecessors. In these, trends toward increase and decrease in the importance of use of the sense can be traced. Although only gross interpretations of behavior are possible, they have particular importance in suggesting the course of habitat occupancy and shifts in the origin of mammals and in various phyletic lines of mammals, including primates. This is developed further in the final section in the paragraphs on the origin of mammals.

The sense of hearing. Both the middle ear and the inner ear of tetrapods are well reflected in skeletal elements. Sequences of changes in the middle ear have been studied in some detail, although only very generally cast in a temporal framework. Interpretation of accompanying changes in hearing is much more difficult and has been the subject of considerable controversy. Most work has been done on the steps leading from reptiles to mammals. Among mammals as well, however, modifications of the middle ear chamber, such as its great enlargement in ricochetal rodents and some South American ungulates of the Tertiary, give considerable basis for interpretation of behavior.

The nervous receptors, of course, reside in the inner ear, both those involved in hearing and in orientation. Although usually well preserved, the osseous structures of the inner ear are difficult to study and provide information only on the bony labyrinth and associated utricular and saccular housings. Development and changes of the lagena–cochlea and cochlea can be traced, mostly in reptiles and mammals in which ossification tends to be fairly extensive. In mammals the form of the cochlea allows fairly extensive analysis and its changes can be plotted in time. Very little of this has been done. A crucial stage of investigation is the transition from reptiles to mammals, but at the most critical levels virtually no information is available from the fossils now known.

RATES AND MODES OF CHANGE IN REPTILES AND MAMMALS

General Patterns

The discussions in this section are limited to reptiles and mammals. These two classes of vertebrates provide the best sources of information and currently are the only groups of vertebrates for which enough is known that some general concepts of rates of change of the nervous system and behavior can be developed. Within these two groups treatment is limited primarily to structures that are important in the broad sweep of evolutionary development. Special developments in structure and behavior, the pineal system in reptiles or the auditory system in ricochetal mammals, for example, are of interest but have not been included in the interest of generality.

Figure 1 provides a resume of the broad changes in the nervous system that can be to some extent documented in the fossil record. The history passes from primitive reptiles through the mammals, including primates and man. The figure forms the framework for the following discussions.

It must be clear that this is in no way a phylogeny but is instead a portrayal of characteristic levels of organization as seen successively in the record. Some of the levels, terminated by brackets where new ones have developed, have continued even to the present time. The major innovations, with the limits of time over which they have developed (numbers 2_i–7_i) and the longer periods during which there has been moderate evolutionary stability in basic characteristics (numbers 1–10) are entered. The latter represent times when adaptive radiation is the principal mode of evolution. During the former, shifts toward occupancy of new life habitat zones were characteristics.

Uncertainties about the periods of time through which innovations have developed remain much the same as when Simpson (1944) proposed the concept of quantum evolution. The dashed lines, used to represent such periods, pass from the latest time in which the innovation is known to be absent in presumed ancestral forms to the time that it is first present. Because this gap is usually relatively short, it may be assumed, if somewhat hesitantly, that innovations have arisen rapidly in evolutionary terms. The general estimates of these times of "rapid" innovation indicate that each has taken place over periods of less than 5 million years. Confusing the issue, however, is the fact that innovations have tended to arise independently in different phyletic lines and not always at the same time.

One nonneurological innovation (3_i) is entered. This represents a particularly critical change related to feeding. Many other morphological modifications have taken place, some certainly of primary importance, but they are less significant to the aims of this study. Throughout the series primary attention has been paid to the brain and, beginning with the true mammals,[1] only the brain is considered.

[1]The true mammals, or therians, as used here are the Pantotheria, Eutheria, and Metatheria. Well-known taxonomic difficulties occur at the reptilian–mammalian transition. Many different schemes of classifications have been proposed. To avoid involvement in these problems an informal classification that is useful in context is used here. True mammals include all living forms except the monotremes. The term "quasi-mammal," adopted from MacIntyre (1967), is used to include the monotremes and a number of extinct groups–eozostrodonts (morganucodonts), triconodonts, symmetronts, docodonts, and multituberculates. All animals that to not belong in these groups but that are "mammal-like" are placed in the reptilian Synapsida, either in the primitive Pelycosauria or the advanced group, Therapsida.

Period	Subdivision	Million years		Evolutionary circumstances		
Quaternary	Recent	0				Stabilization of relative
				Hominids	10	brain size; increase in
	Pleistocene	2				neural mass.
	Pliocene		8	Relative mean brain size		Initiation rapid relative
		12		increase, various lines.		increase in brain size.
				Adaptive patterns of brains mostly	9	assumption of hominid brain
				set. Increase in diversity of relative		organization; time uncertain
	Miocene			brain size. 90% (log) relative increase		from direct evidence.
				of brain function of body size.		
Tertiary		26		Relative mean brain size increase; different rates in different groups.		
			7	Increase in diversity in gross structures and in size.		
	Oligocene	38				
	Eocene	54		7_i Development of "intermediate" to "advanced" levels of brain size and organization; different lines following different pathways, with		
	Paleocene	65	6	different rates of change. Detail mostly lacking.		
	Maestrichian			Archaic placental mammals. Basal mammalian brain level. Some relative increase		
	Campanian			in brain size over time. Beginning development of specialized brains relative		
	Coniacian			to major adaptive behavior.		
	Cenomanian	100	5	Very poor record		
Cretaceous	Albian			5_i placental–marsupial split (based on teeth).		
	Aptian	112		Therian mammals, but brain structure, middle ear not known (inferred on basis of later therian conditions).		
		135				

Jurassic	Upper Middle Lower		4 4_i	Brain filling cranial cavity, attaining general brain–body size ratio of mammals. Accesory elements of lower jaw present (except in multituberculates) but dentary–squamosal articulation developed.
		190		
	Rhaetic–Keuper	?210		"Quasi-mammals" appear and persist through Lower Cretaceous.
Triassic	Muschelkalk	220	3	Advent of brain filling brain case. Brain–body size ratio small, reptilian. Reduced stapes. Reduced nondentary elements of lower jaw, beginning of dentary–squamosal joint. Development of turbinals. Reduction (loss) of parietal foramen. Questionable development of skin glands and vibrissae. Size reduction of overall animal. Development of parallel stocks of very mammal-like reptiles (Therapsids).
	Bunter	225		
			3_i	Beginning of cynodont therapsids. Expansion of zygomatic arch and modification of jaw musculature.
	Upper		2	Primitive therapsid carnivores, radiation of lines. After initiation tendency toward size reduction, decrease in size of parietal foramen, moderate reduction of nondentary elements of lower jaw.
Permian			2_i	Enlargement of floccular fossa.
		240		Beginning of therapsids, relatively large parietal foramen.
	Lower	280	1	Relative brain size characteristic of reptiles. No turbinals or expansion of olfactory parts of brain. Stapes large, nondentary elements of lower jaw large. Pelycosaurian mammal-like reptiles, early members very primitive reptiles.
Carboniferous				

FIG. 1 A chart showing historical sequences of brain and sensory evolution. Numbers 1 through 10 represent successive phases during which there was either stability or slow, regular change. The numbers with the subscripts i (as 2_i) represent times of relatively rapid modification at the beginning or end of more stable periods. They relate to events entered under the same numbers without subscripts. The chart forms the basis for much of the discussion and analysis in the text.

The evolutionary series depicted in Figure 1 is a continuum but the sorts of evidence available prior to and following the origin of the true mammals differ sufficiently that it is convenient to treat the levels separately.

Reptiles to True Mammals

General

Four radiations, 1–4 on Figure 1, occupy a period of about 180 million years: pelycosaurs (1), 50 million years; early therapsids (2), 15 million years; late therapsids (3), 35 million years (total therapsids, 50 million years); quasi-mammals (4), 75–80 million years. It was through this long time span that vertebrates passed from very primitive reptiles to the beginnings of true mammals.

Within these periods, relatively slow to moderate evolutionary change occurred, horotelic in the sense of Simpson (1944). Regular changes took place in the many phyletic lines and there was a great deal of parallelism. More rapid modifications are found in a few highly divergent lines, but this does not appear to have been the case along the general pathway that I am following. Such rapid adaptive change generally was followed by extinction. Evolution was generally progressive in the sense of maintaining moderately constant direction. It is therefore possible (e.g., Olson, 1944) to establish a numerical scale of progress from an initial to a final condition and to determine rates in terms of percentage of attainment of the final condition, here mammalness. In this connection, and in all such averagings that lead to interpretations of rates, however, the appearance of a constant rate, either isometric or allometric, may be merely a function of the materials and methods. Without a closely graded and authentic phyletic series it is not possible to detect variations of rates of change over the times in question. Whenever broad averaging is done, as usually it must be, a concept of steady rates of change arises.

A variety of broad interpretations of general behavioral changes may be made from the data that have contributed to the chart in Figure 1. Development of an increasing capacity for activity, apparently for sustained activity, is most striking. Various bits of evidence (see Olson, 1959; Brink, 1956; Tatarinov, 1967) point to an increasing tendency toward endothermy. Over the full 180 million years, an extensive reorganization of the majority of body systems took place. The pace was generally slow but with long plateaus alternating with more rapid shifts in fundamental organization. If a living insectivore, a hedgehog or tenrec, for example, is compared with a terrestrial lizard, or with *Sphenodon*, the general order of magnitude of change can be approximated.

Pelycosaurs to Therapsids

Two fundamental neurological changes (2_i and 4_i) are indicated on the chart for this period. These were innovative. Other listed modifications appear to have taken place at moderate rates as parts of the continuing adaptive radiations in the several phyletic lines. In addition, the masticatory innovation 3_i is entered because of its great significance to cynodont and eventually to mammalian evolution.

Pelycosaurs evolved for about 50 million years with no evidence of neurological changes. The only strong departure from the norm was in the development of a very large pineal foramen in some divergent phyla. There is no evidence of change in the form or general capacity of the brain case. This conforms well to Jerison's findings (see Figure 3) of the basic differences in relative brain–body sizes between reptiles and mammals. Pelycosaurs fall well within the reptilian minimum convex polygon (see footnote, p. 70). As far as size is concerned, of course, data not on brains but on cranial capacities are compared. Use of cranial capacity as a means of estimating brain size requires an assumption of constant relationship between the capacity and the brain volume. The relationship in mammals is roughly 1:1 but in reptiles it is not. Among living reptiles a fairly constant ratio does exist and this can be extrapolated to many fossil reptiles (Jerison, 1969b). What is being compared, however, is cranial volume and any changes in the ratio of brain volume to cranial volume are likely to be missed. This probably does not apply significantly to the pelycosaurs, but there is reason to believe, as noted below, that there has been such a change in the course of therapsid evolution.

The first evidence of change (2_i on the chart) is the appearance of a deep floccular fossa along with distinct impressions of the ventral and lateral structures of the posterior part of the brain on the osseous brain case. This appears almost simultaneously, as far as the record shows, in three groups of therapsids: gorgonopsians, therocephalians, and anomodonts. It is found in cynodonts, which appear later, and apparently in all advanced therapsids. The three stocks in which it first appears are present in the earliest vertebrate containing Permian beds in South Africa, at a time probably equivalent to the Middle or Lake Kazanian, or possibly earliest Tatarian of the USSR.

A gap in knowledge of the cranial structure exists between the latest pelycosaurs of the sphenacodont line (ancestral to therapsids) and the first appearance of these new features. This is a result partly of the scant record and partly of the fact that cranial structure has not been studied in the few intermediates that are known. The gap is both temporal and morphological. Without question the change in brain structure and relative brain–body size is relatively rapid, taking place over but a few million years, but nothing more can be said about the exact course it has taken or the rate at which it has occurred.

Within the therapsids a rapid basal radiation, beginning early in the Kazanian, produced a number of distinct phyletic lines. As usual after such an initial radiation, a few lines persisted and most became extinct. All of those that persisted had this brain innovation. It may reasonably be associated with development of greater activity and, in particular, with a capacity for sustained activity. It is one of several changes that suggest a beginning of endothermy, perhaps as an integrating physiological process about which many adaptations have grouped (Olson, 1959).

The evolving therapsid lines developed many structures in parallel. Most of the changes represented increasing expression of features now associated with mammals. Among these changes were special modifications related to the sense of smell.

As far as can be told, gorgonopsians, anomodonts, and primitive therocephalians were pelycosaurian in the cerebral parts of the brain and the olfactory apparatus. Their skulls show no evidence of significant changes. Among cynodonts, however, enlargement of the cerebrum and development of complex turbinal elements in the narial chamber occurred. Olfactory bulbs probably were well developed. This is the second significant neurological change (4_i in the chart). Bauriamorphs and advanced therocephalians, other therapsids that are mammalian in many features, give evidence of similar structures, almost certainly developed in parallel. Without much question these changes accompanied an increasing use of and dependence on the sense of smell, relative to a vast series of changes in these increasingly active and more selective predators. It must be noted, however, that once initiated modifications of this type have been maintained and sometimes elaborated in herbivorous stocks that include large, slow-moving animals. The floccular fossa, for example, was developed, apparently independently, in the anomodonts and both the floccular and nasal modifications were present in a branch of cynodonts that led to large herbivores.

Just how the changes of the forebrain and hindbrain were related is not certain. Apparently the former came first. Both appeared in several lines, apparently independently and rather suddenly. Evolution appears to have been rapid as compared to the course of change of most structures of the skull and skeleton. The general therapsid habitus is attained before the cranial modifications appear so, initially at least, they appear to have followed in the wake of other specialization.

In any event, among the cynodonts, where the ancestry of mammals is now generally sought, the cranial cavity becomes increasingly occupied by the brain, producing a condition much like that in quasi-mammals where full occupancy has also occurred. A possible major difference, however, may occur between fairly early cynodonts, such as *Thrinaxodon,* and quasi-mammals represented by triconodonts. The brain–body ratio appears to have changed considerably. Although *Thrinaxodon* and the few other therapsids for which information is available show the ratio to be similar to that of typical reptiles (Jerison, personal communication 1972) that of the triconodont is more comparable to the ratio in primitive mammals, such as the opossum.

Along with the brain modifications, a number of changes of structure related to the nervous system and behavior took place. These appear to have been brought about by relative slow, progressive changes, not by rapid innovation.

The cynodont shift in the masticatory mechanism. This is item 3_i in Figure 1 and is not neurological. The zygomatic arch bowed outward leaving space between it and the occluded lower jaw. This served to accomodate a new, masseteric, subdivision of the adductor musculature. Along with dental modifications and changes in the jaw joint, new ways of procuring and treating food came into being. Barghusen (1968), who worked out and described this change, felt that it was a key to the development of many mammalian features. Among other things, these are related to modifications of the parts of the skull associated with the middle ear and hearing.

Size reduction. The reduction of overall size appears to have been a trend in several therapsid lines. The first mammals were very small, not much more than shrew size. Implications of size reduction are manifold, especially as they relate to problems of endothermy, food sources, and mechanics of food procurement and treatment, and of course to the sense organs and nervous structures as a whole.

No step by step phylogenetic series showing such a reduction has been found. An alternative hypothesis, of course, is that there has been a line of very small cynodonts through which the evolution to the quasi-mammals has passed. If so, they are totally unknown. It is assumed that this has not been the case, both because of the record and because of practical physiological problems in the development of endothermy under such circumstances.

Recognition of this trend as an overriding circumstance is critical in evaluating the trends of morphological change that have led to mammals.

Modifications of neurological and sensory structures. Auditory, narial, pineal, and tactile sensory structures have left some evidence of changes, some well documented and others more obscure. Also in this category is the pituitary, which, as is the pineal apparatus, is part of the brain but a special structure that can be treated separately. All, of course, have bearing on the interpretation of behavior. Along with other modifications of the skeleton, the dentition and masticatory apparatus, and the muscle system, they provide an excellent overview of the transformation from a typically reptilian organism, the pelycosaur, into a mammal.

The middle ear. The story of the reduction of the quadrate and articular bones of the skull and lower jaw, respectively, and their eventual incorporation into the ossicles of the middle ear is well known. Less well understood is what relationship this may have had to modifications in the reception of sound waves. Most of the quasi-mammals had not as yet attained the typical mammalian ear. It has been doubted (Tumarkin, 1955) that early synapsids could hear via a tympanic membrane and ear ossicles. It appears to me, at least, that the modifications that have set the stage for the middle ear of mammals have been consequent upon changes of the masticatory system and indirectly, if at all, related to adaptation to sound reception. By the time the quasi-mammals had come into being the articular and presumably the quadrate were small and a dentary–squamosal joint linked the jaws. The story of evolution of the middle ear from this point to true mammals, and rather late ones at that, is a blank.

The inner ear. Pelycosaurs had a characteristically reptilian lagena–cochlea. The size and depth of the saccular recess in various therapsids suggest that modifications were in progress but preservation is such that it is not known what they were. The shapes of the recesses are different in the various lines and indicate somewhat divergent specializations rather than a course in the direction of mammals. Even the quasi-mammals do not indicate a true mammalian condition, but the evidence in one direction or the other is slight.

Semicircular canals are stable in structure among the tetrapods. One modification occurs during therapsid evolution. The vertical canals rotate relative to the horizontal canal and the horizontal canal changes in orientation relative to the long axis of the skull. These changes, it appears, accompany a downward deflection of the anterior part of the skull so that the long axis is no longer a more or less direct continuation of the axis of the vertebral column. Presumably this is related to developments in food procurement and in the way that the sense organs of the head meet environmental stimuli. The change took place independently in several lines.

The parietal fossa. This structure poses some interesting functional problems during its evolution. Edinger (1955b, 1956) has given it attention and others have dealt with it in passing. In a number of different groups of primitive reptiles and near reptiles, the parietal fossa is inordinately large, based on a scale of comparison with the size of the foramen magnum as suggested by Edinger. The parietal organ, part of the pineal system, seems to have filled the aperature. Enlarged fossae occur mostly in large, slow-moving herbivorous reptiles (for example, *Diadectes,* caseids, and large procolophons) but also are present in a few fairly small edaphosaurian eothyrids, small procolophons, and varanopsids. The last are carnivores. Most of these animals lived in or near water. That the organ was a receptor of external radiation, as in some living reptiles, would seem certain. Many suggestions about its function in the fossils have been made—temperature sensing, control of dessication, actual sight, and so on—but none is on very firm ground.

Among the sphenacodont pelycosaurs, from which the therapsids arose, the parietal fossa was relatively small. Among a number of the primitive therapsids, however, it was very large. Its importance as a sensory organ is emphasized by the fact that it is generally raised above the skull roof, lying in a crater. The fossa was large, but not inordinately so, in phthinosuchids, forms intermediate between sphenacodonts and some of the more advanced therapsids. At this stage the parietal organ undoubtedly had an important function. In the course of evolution of several therapsid lines, including the cynodonts, the size of the fossa became reduced and in some late members it was absent. It was not present in quasi-mammals but in them, as in some true mammals, pineal structures might have been just under the surface and light and temperature sensitive.

The evolution of this organ and associated structures illustrates graphically both the fascination and the frustration of evidence from fossils. There can be no doubt that the pineal apparatus has had an important bearing on the evolution of behavior. To some extent it can be placed in a temporal framework. How it has related to behavior, at best, must be highly speculative, however, for there are no living animals that form even remotely useful analogs.

The hypophyseal fossa. The hypophyseal, or pituitary, fossa is relatively immense in sphenacodont pelycosaurs and gorgonopsians. Its osseous walls laterally and ventrally are well ossified and it may be assumed that the fossa was fully occupied. The pituitary quite certainly was very large. This was not the case in the predecessors of the sphenacodonts or in the cynodonts and therocephalians. Therefore, the pituitary and associated structures seem to have undergone evolutionary enlargement and then marked decrease. This can be cast in a fairly accurate time scale.

As in the case of the parietal organ, however, functional and behavioral interpretations are difficult. Edinger (1942), noting the large pituitary of giant birds and large mammals, suggested it was related to growth and giantism. This does not apply to the mammal-like reptiles. Intriguing suggestions can be made about relationships to changing reproductive habits, to alteration of metabolic rates, and so on. These are completely untestable. Something was going on and possibly in context with many other changes one most plausible explanation will be forthcoming. As yet there is none.

The narial–facila region. The development of the olfactory structures has already been discussed. The narial–facila region, however, gives another bit of information that has bearing on behavior, although little or none on rates of its evolution. In a number of lines of therapsids, bauriamorphs, therocephalians, and some cynodonts, grooves on the external surfaces of the bones of the snout suggest the presence of nerves and blood vessels. These have been interpreted to indicate the presence of vibrissae (Tatarinov, 1967). In this event, in addition to the sensory interpretation, this implies the presence of hair and this has been related to the existence of endothermy (Brink, 1956). The presence of vibrissae is certainly open to question but, in any event, a trend away from the relatively insensitive facial region of reptiles toward a more mobile, sensitive mammalian condition does seem to have been taking place. This was a slow modification and probably associated with an emphasis of the sense of smell and modifications of the masticatory apparatus that widened the potential of food selection and preparation.

The nervous and sensory changes considered together provide a framework within which the course and rates of changes from reptiles to mammals has occurred. Each can be treated separately but it is together that they are most significant. The evolution was long–about 180 millions years–from the beginning to the emergence of true mammals. It was marked by a few major innovations that provided the structural and physiological core around which the parallel changes in adaptive radiations took place. Very little specific information on either behavior or rates emerges. However, a broad picture of the course of major change into which the bits of specific detail from the fossil record and extensive data from living organisms can be fitted does exist.

Quasi-Mammals to True Mammals

The quasi-mammals form a group structurally intermediate between therapsids and true mammals. They are known mostly from teeth and jaws and appear to have radiated from an ancestral type with tritubercular teeth into several divergent lines. Of these only one led to the true mammals. All quasi-mammals were small and all except the multituberculates were omnivorous or insectivorous. The multituberculates were quite apart from the others. They were herbivorous and some late members were relatively large. They appeared in the Jurassic and lasted into the Eocene, whereas the other groups ranged from the Triassic into the middle part of the Cretaceous, covering a period of 75–80 million years.

At the upper end of their range, in the Mid-Cretaceous, the quasi-mammals overlapped the true mammals in time. For the whole period of time only a small number of skulls is known and few of these have given any information on the nervous system. Data have come primarily from the following specimens: a triconodont (Simpson, 1937; Kermack, 1963); a primitive triconodont-like form, *Sinoconodon* (Patterson & Olson, 1961); a pantothere, *Amblotherium* (Edinger, 1964a); and a multituberculate, *Ptilodus* (Simpson, 1937). In addition, skulls and fragments of *Morganucodon* (or *Eozostrodon*) have given some information about the cranial structure of this earliest and most primitive of the quasi-mammals.

An important conclusion that comes from this scant evidence is that the cranial structure appears to be very similar in all of them. Edinger (1964a), commenting on this, concluded: "In extremely macrosmatic mammals of 180 to 140 million years ago, an expanded but still palaeencephalic cerebrum overlapped the midbrain which was still of reptilian type as were the hindbrain and other characters in these most ancient mammals." Edinger, as well as Kermack (1963), concluded that the brain case and presumably the brain was typically reptilian. More than in most therapsids, however, the brain appears to have filled the cranial cavity nearly completely. Although the general form was not greatly different, as revealed by cranial casts, there does appear to have been the marked difference in relative brain–body size between *Thrinaxodon* and the triconodont, as noted above. It cannot be known how representative these creatures are, for there may have been great variability within the groups they represent. What little data exist suggest that they are more or less representative.

The therapsid type of brain persisted over many millions of years. It first arose among carnivores and persisted as size decreased in descendant omnivores and insectivores. It was retained as well in herbivorous lines, among the cynodonts in particular, in which size either was maintained or increased.

As the archosaurian and lepidosaurian reptiles underwent major radiations beginning in the Triassic and continuing through the Cretaceous, the previously dominant therapsids decreased sharply, being preserved only in their quasi--mammal descendants. A strong behavioral shift was associated with the evolution, with a strong restriction in adaptative zones compared with those of the therapsids.

In a reply to a letter by Nyberg (1971), Jerison (1971b) put forth some suggestions about this new life zone and its relationships to the evolution of the brain. Taking into account the pertinent information of therapsids and quasi-mammals, data on primitive true mammals, and the relevant detail from living mammals, he concluded that therapsids shifted from a diurnal to a noctural existence. This he related to development of new capacities for distance sensing, particularly smell and hearing, which made it possible for the animals to function in the nocturnal world. Development of these senses was related to increase in brain size. He analyzed the consequences of these changes with respect to conditions in descendant mammals. Developing endothermy was integrated into the adaptation to nocturnal existence. In essence, he sums up the behavioral basis for the later diurnal radiation of true mammals as one established while they occupied nocturnal habitats, which made possible survival and developed "preadaptations" for the adaptive evolution seen among more advanced mammals.

If this interpretation is accepted, and its appears plausible, some interesting points arise. First, the basic enlargement of the olfactory and acoustical regions of the brain developed in fairly large, successful, and presumably diurnal therapsids. Also, there is at least some evidence that these creatures were endotherms. They appear to be themselves "preadapted" for the nocturnal existence that Jerison envisages. Passage toward the new adaptive zone may have provided strong selective pressures toward decrease in size along with the competition with increasingly numerous and diversifying archosaurcian and lepidosaurian reptiles. With this also came strong selection for modified diets, away from large prey and the tearing away of flesh, to small prey, insects and other invertebrates, and oral preparation of food.

Filling of the cranial cavity by the brain may be in part a function of decrease in size of the animals. However, it also tended to occur in some gorgonopsians, therocephalians, and anomodonts, which were moderately large. At least the first steps of the transition were therefore realized prior to the reduction of size, perhaps somehow related to changing biomechanics of the skull and jaws. Increase in relative brain size, leading to essentially true mammal proportions, appears to have taken place along with the transition to smaller, perhaps nocturnal animals.

The rate of evolution involved in the shift from one adaptive zone to another often appears to have been very rapid. How much of this is real and how much is a matter of the record is uncertain. In this instance, where a marked shift clearly occurs, even if its precise nature be questioned, the rate appears to fall well within the horotelic range characteristic of vertebrate evolution as a whole. The stage presumably was set by the two basic brain modifications among therapsids and then by radiations that carried the therapsids into many life circumstances. The basis for the shift, which had great consequences in the origin of mammals, had appeared during the Middle Permian but was fully realized through typically slow progression only by the Late Triassic, some 40 million years later. The part of the radiation among the cynodonts, which produced very small, probably nocturnal animals, was the most successful of the many courses followed.

Very little is known of the transition from quasi-mammals to the true mammals, and this is particularly true as far as the brain is concerned. From the very scant evidence, it appears that no major modifications have taken place, but this is based on only a most general idea of the shape of the cranial cavity for quasi-mammals and estimates of brain–body size relationships from extremely scant information. Both the extent of the evidence and the extent of the extrapolation that have gone into the estimates must be kept strongly in mind.

TRUE MAMMALS

Perspective

Study of the nervous system of the true mammals is dominated by evidence from endocranial casts as far as fossils are concerned. These, with accuracy varying, reflect gross features of the external anatomy of the brain and offer a basis for estimating brain volume. Most of the work has been directed along two avenues–analysis of brain structure and analysis of brain size. The two are complementary and overlapping but they have been pursued somewhat independently.

Edinger has paid particular attention to structure and, most recently, Jerison, to size. Both types of study, of course, depend heavily on the extensive work by a large roster of students of structure and relative brain size in living animals. Both Edinger and Jerison have built upon this work as well as on the foundations laid by many earlier students of fossil brains.

For all that has been done, the data needed for generalization and for coordination of studies of living and fossil materials is far from sufficient. Information from fossils is difficult to obtain and even more difficult to interpret. Much of what is potentially available has not been developed and studies. If this area of study is to thrive, the same intense efforts that have been devoted to the study of evolution of dentitions must be applied to fossil endocasts. Even if materials are exploited to their fullest, there still will remain many morphological and temporal gaps. Gradually many of these may be filled.

Work to date has dealt in fair detail with phyletic lines among the Equidae, Camelidae, and Hominidae. Attention has been paid to several kinds of archaic mammals and, in a spotty fashion, taking advantage of findings as they have been made, to a fairly wide range of genera scattered through the placental orders. From these studies have emerged some generalizations that concern rates of physical evolution of the brain and their implications with respect to rates of evolution of behavior. These form the basis for discussion in this section. The framework in which they can be considered is that given in the Tertiary and Pleistocene portions of Figure 1.

The most archaic known true mammalian brain, as determined from primitive placentals and marsupials, including opossums and some primitive insectivores, appears to be little advanced over that found in quasi-mammals. This is in some contrast to the ideas of Edinger (1964a) in which the contrast to what she terms the "palaeencephalic brain" of quasi-mammals and neencephalic brain of true mammals is given as follows: ". . . the midbrain had emerged to the brain surface. Now it was stretched between a cerebrum that had acquired a distinct neocortex and a fully mammalian cerebellum. Neencephalic formations had been added to its walls and acoustic colliculi to its roof."

Edinger did not consider relative brain–body sizes and this led her to notable differences with Jerison, who pointed them up as distinct differences between therapsids and quasi-mammals and as strong points of similarity between the latter and true mammals. Interpretations of any but the most gross structures of the brains from the early fossils must be considered largely guesswork.

The level of behavior of animals possessing the primitive brain evisaged by Jerison in quasi-mammals and early true mammals, or by Edinger in the latter only, probably compared broadly with that exhibited by opossums and primitive insectivores. Just when it came into being is uncertain. If it was present in the quasi-mammals, as suggested by Jerison, then it was developed no later than the Mid-Jurassic and possibly as early as very Late Triassic. If it should prove with additional evidence that there were notable changes between quasi-mammals and true mammals, as Edinger and also Kermack (1963) have suggested, then the date is much less certain.

In the first instance, with the appearance no later than Mid-Jurassic, the radiation of true mammals began long after the basic brain type had been acquired. In the second instance, however, this is not clear. The variety of mammalian types present even in the Early Paleocene and the rapidly advances in adaptive types during the rest of the epoch indicate the beginning of an extensive radiation no later than the very Late Cretaceous. This gives evidence of a rapid passage across some organizational threshold to the full attainment of mammalian structure and behavior by the Late Cretaceous. The radiation could not have taken place without attainment of the type of brain found in the primitive true mammals. However, that attainment of this condition has coincided temporally with the beginning of the radiation cannot be demonstrated on the basis of known materials.

The archaic mammalian brain was ancestral to all more advanced brains of subsequent true mammals. It has been found in all primitive groups in which study has been possible–condylarths, arctocyonid creodonts, insectivores and primitive marsupials. If there has been such a common ancestral brain type, as the evidence strongly suggests, it may be assumed to have been present in the ancestral base of all true mammals. This base certainly was well established by the end of the Late Cretaceous and, on the basis of teeth, has been recognized as far back as the end of the Lower Cretaceous. If the latter interpretation is accepted, then, as in the case of the quasi-mammals, there is a long period over which the archaic true mammalian brain has existed with very little modification. In view of various other rates of change in brain patterns this seems at least plausible.

Marked ecological changes occurred during the Late Cretaceous. Angiosperms became the dominant plants and there were marked changes in reptilian faunas, with extinctions at the end of the Cretaceous. By the end of the Cretaceous new opportunities existed for the radiation by animals at the true mammalian grade of organization. Such animals may have existed at this level for a long period of time prior to the appearance of circumstances suitable for their radiation, instead of having passed an organizational threshold and rapidly deploying.

Some phyletic lines of brain development from the primitive to more advanced states can be traced in equids, camelids, hominids, and creodont carnivores. These indicate that each line of development has been marked by brain structures specialized in accomodation to the particular modes of existence. Brain specialization reflects the selective pressures under which evolution in different phyla is occurring. Brains in this regard appear to behave no differently than do other systems under selection. The matter of brain specialization in evolution has been stressed by Edinger (1961). As work on racoons by Welker and Seidenstein (1959) has shown, detailed correlation between enlargement of particular parts of the brain and behavior occurs and is readily detectable. Preliminary studies of some fossils have shown that this can be applied to fossil endocranial casts where enough information on moderns is present.

Answers to some of the general problems of rates of behavior evolution seem to be approachable if lines can be followed from the primitive base, which appears to be common to all true mammals, into more specialized lines. This can be done if, and it is a big if, enough detail of the brain surface can be determined to allow study of the nature and rates of change of areas that can be related to actions of particular parts of the body. To do this at present is out of the question, because of lack of fossil materials, of appropriate phylogenetic knowledge, and of detailed mapping of appropriate living animals.

In some lines of true mammals characteristic brain patterns of particular groups appeared after the basic structural adaptations of other parts of the body had come into being. This is an important generalization. Stated otherwise it means that brain structure has not necessarily been a key innovation as a new adaptive zone has been occupied but has come under selective pressures after the advent of that occupancy. This contrasts with the apparent "preadaptation" of the therapsid brain for a new mode of life noted above.

How general this phenomenon may have been among mammals remains uncertain. Problems of phylogeny and identification of key genera have brought out arguments in the cases in which it seems to have been well documented. It was emphasized in the work of Edinger (1948) on equids and appears in camelid evolution (Edinger, 1964b, Jerison, 1971a). It seems to be the case in the hominid sequence of *Ramapithecus–Australopithecus africanus–Australopithecus habilis.* In the earliest equids and apparently in the early camelids the brain pattern, even after the development of recognizable phyletic traits in other characters, still had features of the basic mammalian type found in condylarths.

If this generalization holds, of course, it has important behavioral implications, indicating that behavior related to feeding, locomotion, and defense, among other activities, has not been conditioned by preadaptive or commensurate evolution of the central nervous system. Only with their establishment did selection act on the central nervous system to give it the characteristics related to special functions of the animals.

The general features of the equid line were established in the Eocene, in the lineage of *Hyracotherium, Orohippus,* and *Epihippus.* The extremely primitive brain attributed to *Hyracotherium* by Edinger has been questioned and published evidence on brain casts for this series is slight and identifications are somewhat uncertain. It does appear, however, that between the Mid-Eocene *Orohippus* and the Lower Miocene *Mesohippus* marked changes occurred and that the typical equid brain type came into being.

The same sort of timing appears to be found in the Upper Eocene to Lower Oligocene evolution of the camels. Here, however, there exist some problems as to ancestry and the exact nature of the brain configuration. In each of these lines, once the shift to the brain type characteristic of the particular group had taken place, regular, progressive changes occurred, with the size increase of the brain intimately related to increase in body size.

Several successive levels of brain evolution occurred among the true mammals during the Tertiary. This follows mostly from studies made by Jerison. In arriving at this generalization brain evolution has been viewed in the perspective of grades, rather than of discrete phyletic lines, or clades. Concern is less with occupancy of niches, as in the generalizations to this point, and more with statistical means and variances based primarily on cranial capacities and their relationships to body size.

Before this generalization is examined further, some note must be made of the differences of opinion regarding the utilization of cranial capacity or brain size as a criterion in assessing behavior or function. This has occupied the attention of both neo- and paleoneurologists. The great advantage of using size criteria is that they supply immense amounts of data, from both modern and fossil animals. Jerison (see especially 1970a, b) has been the most vigorous and active proponent of the use of these data in recent times. He, as have others, has treated cranial capacity as representative of brain size or weight, as must be done with fossils, and has used body weight or volume, either determined directly on modern animals or by conversion of linear dimensions in fossils. Many ways of expressing the relationships between relative brain and body size have been devised. They are discussed in detail by Jerison (1970b, 1973). In the illustrations cited below, Jerison has used what he feels to be the most effective way of representation.[2]

A problem of using volumetric data lies in the interpretation. Intuitively it seems reasonable that major differences in brain volume would be accompanied by different behavior patterns. In general larger brains should have a greater capacity to process information. However, this is not necessarily the case except when volumetric differences are very great. Given the same basic organization, brains of quite different capacities operate with equal effectiveness. The crux of the situation rests, it appears, on the degree of similarity of organization. If it is assumed that within a group brain organization is about the same, then major differences in size may have some importance. Another factor enters in, however: the size of the brain relative to body size. If the brain is considered isomorphic throughout the placentals, then the brain of a small animal mapped over to a large animal should be much the same size as that of the large animal. As the equation $E = 0.12P^{2/3}$ shows, however, this is not strictly the case. Jerison has indicated that about 90% of the variance of the log of brain size is accountable for by the log of body size. Therefore, although there is constancy in relationships, there is also unexplained variance.

Such students as Jerison, who rely to some extent on relative brain size for understanding behavior, find significance both in the common aspects of the brain–body relationship in the true mammals and also in the departures of the relationship from the expected values. Although not denying a gross size factor some importance, particularly as it reflects basic differences in organization,

[2] The problems faced here are several. First, it must be established whether in fact there is a consistent body–brain size relationship with the group being considered here, mammals, but more broadly vertebrates as a whole. It has repeatedly been shown that such a general relationship does exist. The power function $^2/_3$ gives a good approximation and various indices using it have been developed. Given this, there is then the problem of establishing a basis for comparing different populations. Jerison has used an "expected" brain–body size relationship, making use of the equation $E = 0.12P^{2/3}$, where E is brain size and P is body size. With data from a large suite of living mammals the regression, a straight line on a double logarithmic plot, can be plotted and, of course, the scatter values around the line entered to show variance. The line may be taken as an expression of expected values. Departure of individual ratios or those of other groups, subsets of the modern sample or groups of fossils, may be entered and compared as desired. Comparisons may be either graphical or statistical.

In lieu of using an index, Jerison has employed a technique involving minimum convex polygons, plotted as explained in his various papers (see, e.g., Jerison, 1970b). These provide a graphic representation and allow for easy visual comparisons of populations. In effect, what is developed is a bivariate space that encloses all the ratio points of relative brain–body size for a designated population. Other spaces may be similarly plotted and compared. Variance is not automatically entered or readable but it can be included if desired and sometimes this is done. These two methods are illustrated in Figures 2 and 3.

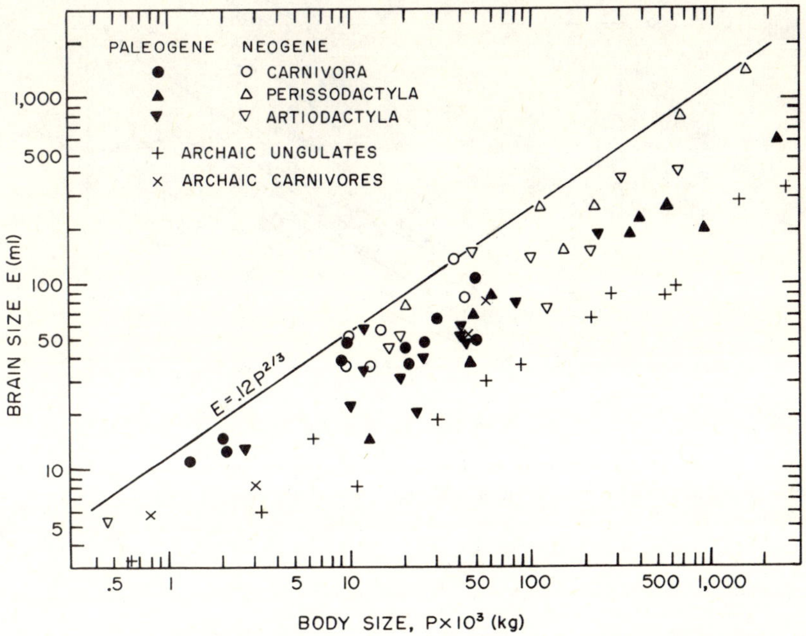

FIG. 2 Brain size (endocast volume) as a function of body size in 69 fossil ungulates and carnivores, logarithmic scale. The line is fitted to a large suite of species of living mammals on the basis of the equation $E = 0.12P^{2/3}$, where E and P are brain and body size, respectively. (Redrawn from Jerison, 1970a with the author's permission.)

others argue that organization is more important than size. Edinger has usually followed this general pathway. Holloway (1968) has taken an extreme position, as follows: "This aspect of neural organization . . . should dictate clearly that appreciation of function on the basis of cranial capacity is almost nonsensical."

Recognizing that these differences of opinion exist, we may return to the generalization. Figures 2, 3, and 4, after Jerison, collate a great deal of information and form the basis for the generalization about the levels of brain evolution during the Tertiary. The diagrams are based on relative brain–body size as discussed in Footnote 2. What they show is that if representative samples are taken from selected, successive geological time units, there emerges a series of stages in which the values in the groups fall successively farther from the expected values determined from modern animals going back in time. Both methods of graphing in Figures 2 and 3, respectively, show this. It is also clear in the size–frequency distributions in Figure 4. If the means alone are used, as in Figure 5 (based on data from Figure 4), it is possible to plot and express in suitable form the rate of change of the mean with time.

A second aspect, however, also appears and has been considered by Jerison. As seen in Figure 4, with the increase in the mean of relative brain–body size the spread, or variance, around the mean increases. At each grade level, marked

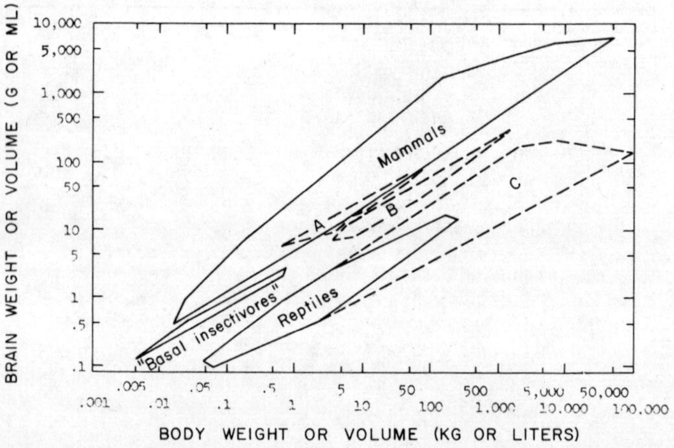

FIG. 3 Minimum convex polygons for living and fossil mammals and reptiles. (Redrawn from Jerison, 1970b; based on a technique described in that reference with author's permission.)

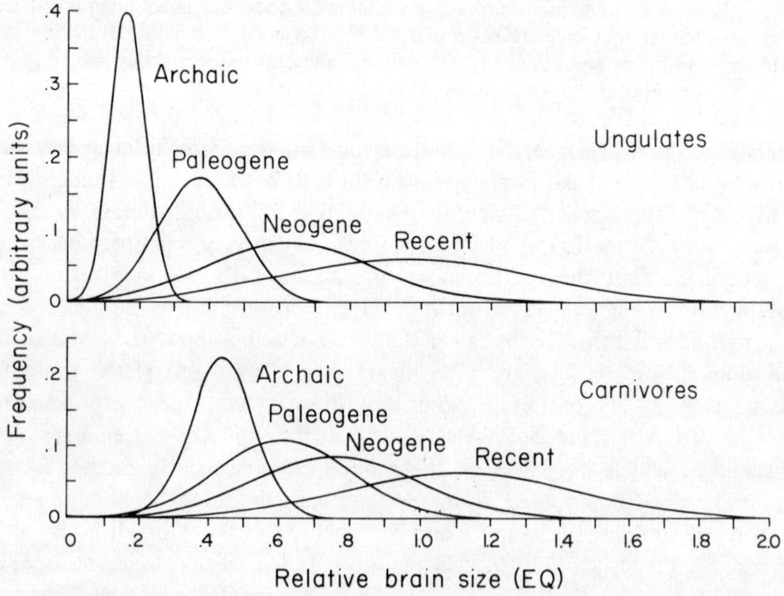

FIG. 4 Changing distributions of relative brain size as the brain evolved. Means and variances are based on cumulative frequency distributions of relative brain size as measured by the encephalization quotient (*EQ*) in fossil and recent carnivores and ungulates. $EQ_i = E_i/0.12P^{2/3}$ where E_i is brain size and P is body size. (Redrawn from Jerison, 1970a with author's permission.)

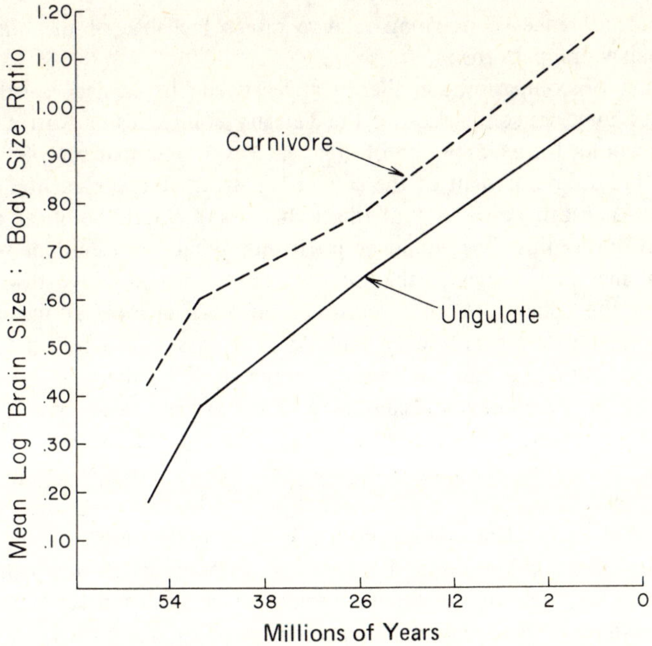

FIG. 5 The means of relative (log) brain sizes as shown in the frequency distributions of Figure 4 entered on a time scale. (Data from Jerison, 1970a.)

(following Jerison) by a greater capacity to process information, the spread of variation indicates a widening adaptive radiation. With this radiation, neurological structures become increasingly varied as they become specilized for different life conditions, and the variation is reflected in the relative brain–body size distribution.

Use of brain size data in assessing evolutionary changes appears to point up several aspects of brain evolution not otherwise evident. The progression of relative brain–body size and the increasing variation just noted are two. These provide ways of analyzing rates of change. The relationships of means and variance and of expected versus determined values point up relationships between concepts of grade levels and special niche occupancy and indicate a need for joint consideration of the two in assessing the course and rate of changes.

Another area of study indicated in size data but largely undeveloped is the comparisons of different populations with respect to levels of development. As suggested in Figure 4, and as noted by Jerison in his discussions of it, carnivores and herbivores from commensurate levels within the fourfold grouping show different means and variances. Different segments of evolving faunal complexes may be compared on this basis and it appears the comparisons can lead to significant inferences concerning relative rates of change of the brain and of related behavioral patterns.

This sort of comparison, in theory at least, can be carried to other faunal complexes. A gross comparison of the Tertiary complexes of North and South America would, for example, be of great interest. A comparison of the relative levels of brain development of the carnivore marsupials and placental herbivores of South American Tertiary or of the herbivores of South American and North American during this time would be intriguing. What has been done with North American mammals indicates that such a venture would be worthwhile. However, here the usual bane of paleontological work arises. For the most part, necessary materials are not to be had. As time passes, new collections help to improve the situation and particularly in recent years there have been major advances. Even were adequate samples available the immensity of gathering data is forbidding. These problems, plus the ever-present difficulty of interpreting meanings, militate against much early major success in this potentially fascinating area of study.

The several generalizations as noted in this section are valuable and give insights to brain evolution and its relationships to rates and behavior not available without use of the fossil record. Work on other parts of the nervous sensory systems of mammals has some chances of success but currently has not yielded much that can be cast in a time continuum for a study of rates of change.

RETROSPECT

What has been presented in this chapter supports the opening statement that the problems of rates of evolution and the nervous system and behavior remain refractory. An attempt has been made to summarize the present state of knowledge without a tedious review of the history of the study or a point by point presentation of the work that has been done. A fairly coherent picture of the broad sweep of evolution, with emphasis where possible on rates, has emerged. Nowhere has it been possible to derive precise rates of change for nervous structures or functions. The threads connecting osteological remains to nervous structures and these structures to behavior remain relatively few and weak. More data can be had and more analyses are possible. This is all to the good.

It seems evident, however, that a major problem is that the study of paleoneurology is still in the state of groping, of trying to formulate questions that can be meaningful to neoneurology, and of probing opportunistically without any real sense of direction. One reason is that very few paleontologists have entered into this field more than sporadically. A lack of articulation of the efforts of students of living mammals with those of students of fossils exists, and directions of interest do not seem to coincide closely. Probably one of the most persistent problems is explication of just what the common problems are.

In the face of this conceptual poverty considerable progress has been made. Some generalizations have emerged and these can become focal points for pointed questions and analyses. This is a first step. Paleoneurologists have been trying to make their points of view clear to their neoneurologist colleagues. Tilly Edinger's work frequently had the education of her colleagues as its primary thrust. Jerison brings a neoneurologist's background to a study of fossils. However, consistent direction seems lacking.

As far as increased knowledge of rates of evolution of some parts of the nervous system are concerned, data from the fossil record can make, and have made, contributions. These involve both direct studies of its materials and suggestions, on the basis of the materials, of ways for more accurately assessing rates of development of successive stages seen in samples of the "phylogenetic" series among living animals. Much more can be done along these lines and it seems probable that concentration in this direction can establish some understanding of rates of change and, perhaps, some insights into their causes.

The field of behavior poses much greater difficulties than that of the nervous system itself. At the general level of treatment used by Romer (1959) the fossil record can make contributions. Such speculations as those by Jerison on the nocturnal habitat of late therapsids and early mammals are interesting and point the way for searches for relevant data. Use of the fossil record to indicate the nature of natural selection in actual evolutionary series may give clues to behavior in ancestral stocks. All such studies, although useful, are basically unsatisfying because by and large their results are not capable of being tested. The gulf between the aims of behaviorists and the information available from fossils appear to me to be great and currently, at least largely unbridgeable.

REFERENCES

Barghusen, H. The lower jaw of cynodonts (Reptilia, Therapsida) and the evolutionary origin of mammal-like adductor jaw adductor musculature. *Postilla,* 1968, No. 116, 1–49.

Brink, A. S. Speculations on some advanced mammalian characteristics in the higher mammal-like reptiles. *Palaeontologia Africana,* 1956, **4,** 77–96.

Denison, R. A review of the habitat of the earliest vertebrates. *Fieldiana: Geology,* 1965, **11,** 359–457.

Denison, R. The origin of the lateral line sensory system. *American Zoologist,* 1966, **6,** 369–370.

Edinger, T. The pituitary body in giant animals, fossil and living: a survey and a suggestion. *Quarterly Reviews in Biology,* 1942, **17,** 31045.

Edinger, T. Evolution of the horse brain. *Geology Society of America, Memoirs,* 1948, **25,** 174 pp.

Edinger, T. Hearing and smell in cetacean history. *Monatsschrifte für Psychiatrie, und Neurologie,* 1955, **129,** 37–58. (a)

Edinger, T. The size of the parietal foramen and organ in reptiles. A rectification. *Bulletin of the Museum of Comparative Zoology,* Harvard 1955, **114,** 3–34. (b)

Edinger, T. Paired pineal organs. In J. A. Kappars (Ed.), *Progress in neurobiology. Proceedings of the First International Meeting of Neurobiologists.* Amsterdam: Elsevier, 1956. Pp. 121–129.

Edinger, T. Fossil brains reflect specialized behavior. *World Neurology,* 1961, **2,** 934–941.

Edinger, T. Midbrain exposure and overlap in mammals. *American Zoologist,* 1964, 4, 5–19. (a)

Edinger, T. Recent advances in paleoneurology. In W. Bargmann & J. P. Schade (Eds.), *Topics in basic neurology, Progress in brain research,* Vol. 6. Amsterdam: Elsevier, 1964. Pp. 147–160. (b)

Edinger, T. Brains from 40 million years of camelid history. In R. Hassler & H. Stephan (Eds.), *Evolution of the forebrain.* Stuttgart: Georg Thieme Berlag, 1966. Pp. 153–161.

Edinger, T. & Kitts, D. The foramen ovale. *Evolution,* 1954, **8,** 389–403.

Haldane, J. B. S. Suggestions as to quantitative measurement of rates of evolution. *Evolution,* 1949, **3,** 51–56.

Holloway, R. L., Jr. Cranial capacity and the evolution of the human brain. In M. F. A. Montagu (Ed.), *Culture: Man's adaptive dimension.* Oxford Press, Oxford, 1968. Pp. 170–196.

Jerison, H. J. Interpreting the evolution of the brain. *Human Biology,* 1963, **35,** 263–291.

Jerison, H. J. Brain evolution and *Archaeopteryx. Nature* (London), 1968, **219,** 1381–1382.

Jerison, H. J. Fossil brains and the evolution of behavior. *Proceedings of the 77th Annual Convention of the Association of Physical Anthropologists,* 1969, 7–8. (a)

Jerison, H. J. Brain evolution and dinosaur brains. *American Naturalist,* 1969, **103,** 575–588. (b)

Jerison, H. J. Brain evolution: New light on old principles. *Science,* 1970, **170,** 1224–1225. (a)

Jerison, H. J. Gross brain indices and the analysis of fossil endocasts. In C. R. Hoback & W. Montagna (Eds.), *The primate brain, Advances in primatology,* Vol. 1. Appleton, New York: 1970. Pp. 225–244. (b)

Jerison, H. J. Quantitative analysis of the evolution of the camelid brain. *American Naturalist,* 1971, **105,** 227–239. (a)

Jerison, H. J. More on why birds and mammals have big brains. *American Naturalist,* 1971, **105,** 185–189. (b)

Jerison, H. J. *Evolution of the brain and intelligence.* New York: Academic Press; 1973.

Kelly, D. E. Pineal organs: photoreception, secretion and development. *American Scientist,* 1962, **50,** 597–625.

Kermack, K. A. The cranial structure of the triconodonts. *Philosophical Transactions of the Royal Society of London* (Series B), 1963, **246,** 83–103.

Kurtén, B. Rates of evolution in fossil mammals. *Symposia in Quantitative Biology,* 1959, **24,** 205–215.

MacIntyre, G. T. Foramen pseudoövale and the quasi-mammals. *Evolution,* 1967, **21,** 834–841.

Nyberg, D. A hypothesis concerning the larger brains of homoiothermy. *American Naturalist,* 1971, **105,** 183–185.

Ohno, S. *Evolution by gene duplication.* New York: Springer-Verlag, 1970.

Olson, E. C. Cranial foramina of North American beavers. *Journal of Paleontology,* 1940, **14,** 495–501.

Olson, E. C. *Origin of mammals based upon the cranial morphology of therapsid suborders.* Geological Society of America, Special Paper No. 55. 1944.

Olson, E. C. The evolution of mammalian characters. *Evolution,* 1959, **13,** 344–353.

Patterson, B., & Olson, E. C. A triconodontid mammal from the Triassic of Yunnan. International Colloquium on Evolution of Lower and Unspecialized Mammals. *Koninklijke Vlaamse Academie Wetenschappen, Letteren en Schone Kunsten van Belgie, Pt. 1.* G. Vanderbroek (Ed.). Brussels, 1961. Pp. 129–191.

Pilbeam, D. *The ascent of man.* New York: Macmillan, 1972.

Radinsky, L. Relative brain size: a new measure. *Science,* 1967, **155,** 336–338.

Romer, A. S. Phylogeny and behavior with special reference to vertebrate evolution. In A. Roe & G. G. Simpson (Eds.), *Behavior and evolution.* New Haven: Yale University Press, 1959. Pp. 48–75.

Simons, E. L. The early relatives of man. *Scientific American,* 1964, **211,** 50–62.

Simpson, G. G. Skull structure of the Multituberculata. *Bulletin of the American Museum of Natural History,* 1937, **73,** 727–763.

Simpson, G. G. *Tempo and mode in evolution.* New York: Columbia University Press, 1944.

Simpson, G. G. *The major features of evolution.* New York: Columbia University Press, 1953.

Stebbins, R. C., & Eakin, R. M. The role of the "third eye" in reptilian behavior. *American Museum Novitates,* 1958, No. 1870.

Steyn, W. Some epithalmic organs, the subcommissural organ and their possible relation to vertebrate emergence on dry land. *South African Journal of Science,* 1961, **57,** 283–287.

Tatarinov, L. P. Development of a system of labial (vibrissial) blood vessels and nerves in the theriodonts. *Paleontology Journal, Academy of Science, USSR,* 1967, **1,** 3–17.

Tumarkin, A. On the evolution of the auditory conducting apparatus: A new theory based on functional considerations. *Evolution,* 1955, **9,** 221–243.

Welker, W. I., & Seidenstein, S. Somatic sensory representation in the cerebral cortex of the racoon (*Procyon lotor*). *Journal of Comparative Neurology,* 1959, **111,** 469–502.

5

Genetic Considerations in Studies of the Evolution of the Nervous System and Behavior

Glayde Whitney

Florida State University

Evolution in the broad sweep of phylogeny can be comprehended with absolutely no knowledge of the genetic processes by which evolution occurs. Indeed, Darwin's own works contain no reference to the contemporaneous discoveries of Mendel. However, an understanding of evolution in process, a comprehension of what the microevolutionary events may be that *en masse* constitute the macroevolutionary happenings of phylogeny, necessitates some understanding of genetics. It has often been pointed out that the major shortcoming of the original version of Darwinian evolution is directly related to a lack of genetical insight. The modern theory of evolution, often referred to as the "neo-Darwinian" or the "synthetic" theory, consists of a synthesis of basic genetics with Darwinian natural selection. Genetics provides the causal mechanisms through which natural selection operates. At the very least a consideration of some of the genetic processes in evolution can be valuable in formulating questions and seeking answers concerning the evolution of the nervous system and behavior, for even though a particular "causal analysis of evolution grounded in genetics may well be wrong, one flying in the face of the established principles of genetics cannot possibly be right" (Dobzhansky, 1967, p. 1).

POPULATIONAL APPROACH

Central to genetic considerations is the realization that evolution is a phenomenon of populations rather than individuals. The concern must be with the variability among populations and among individuals within each population, rather than with a concentration entirely on description of individual types that are taken to represent a group. Darwin realized that variation among the individuals that constitute a population is central to evolutionary considerations. However, typological conceptualizations tend to persist that are difficult to reconcile with an understanding of the genetic processes of evolution. Individual genotypes develop the phenotypes that reproduce to a greater or lesser extent and, therefore, contribute to evolutionary change. However, genotypes (and phenotypes) are most emphatically not what are transmitted across generations.

The source of continuity and of change across evolutionary time is samplings of genes. For each generation the genes of a sexually reproducing local population are reshuffled into new and usually unique genotypes. Individual genotypes are ephemeral single-generation combinations of the genes of a population. Instead of individuals, therefore, it is the interbreeding population as a whole, the gene pool considered *en masse,* that evolves. However, from an interest in behavioral and nervous system phenotypes to a concentration on individual phenotypes, and from there to a typological conceptualization of species, orders, grades, and clades, is all too facile a conceptual trap–a virtual morass of potential misunderstanding.

The typological trap is avoided when the genetic basis of evolution is considered. Mayr (1963) has suggested that

> The replacement of typological thinking by population thinking is perhaps the greatest conceptual revolution that has taken place in biology. Many of the basic concepts of the synthetic theory, such as that of natural selection and that of the population, are meaningless for the typologist. Virtually every major controversy in the field of evolution has been between a typologist and populationist. Even Darwin, who was more responsible than anyone else for the introduction of population thinking into biology, often slipped back into typological thinking, for instance in his discussions on varieties and species [pp. 5–6].

More recently, Dobzhansky (1967) has noted that he agrees with the above quote, his "only cavil is that Mayr's 'has taken place' is over optimistic; I do believe, however, that this conceptual revolution is under way [p. 2]." It is obvious that at the present time considerations of the evolution of the nervous system and behavior are not immune from this source of dissension, misunderstanding, and lack of communication.

GENETICS OF NATURAL POPULATIONS

Organic evolution is a population phenomenon consisting of changes in the frequencies of genes that constitute the gene pools of local populations. For orientation, therefore, one of the first descriptive questions that can be asked has to do with the genetic structuring of local populations or of species. Just how much genetic variability is there in extant populations? Because on a cursory inspection of the world many animal species appear to consist of individuals that closely resemble one another, one popular position in the past has been that in fact there is little genetic variation within most naturally occurring populations. Until quite recently the belief has been widespread that natural populations are composed of groups of individuals that are quite similar to one another genetically. That is, populations have been viewed typologically as being fixed for "wild-type" alleles at most gene loci. According to this view, there existed a few very rare mutant genes with generally deleterious effects. Additionally, genetic polymorphisms [where two or more alternative alleles exist in a population at frequencies higher than those attributable to recurrent mutation (Ford, 1971; Merrell, 1969)], although known to exist, were thought to be infrequent.

Obviously, this view was compatible with some of the classic theoretical illustrations of the impact of natural selection. For instance, if there exists a difference in the Darwinian fitness (relative reproductive performance) of individuals carrying alternative alleles, it can be demonstrated that this selective differential maintained over significant lengths of time can have a profound effect on gene frequency even though the difference in fitness of alternative individuals is very slight. It follows that an infintesimally small selective advantage, maintained over a sufficiently great number of generations, can drive a particular allele to fixation in the population. Such as allele would be the common "wild-type" gene. Therefore, one of the results of such an exercise in the algebra of population genetics was the explanation of the common prevalence of homozygous wild-type alleles.

The alternative to this classical view is that genetic variation is much more prevalent than previously recognized. Instead of fixed wild-type alleles at most loci with alternative alleles existing at only a small proportion of loci within any single population, proponents of this second view suggested that polymorphic loci were much more common; that series of multiple alleles could be maintained at many loci in natural populations of sexually reproducing species. The general evidence in support of this view also came from observations of the phenotypes that existed in natural populations and from the results of genetic experiments. The observational evidence was simply that virtually continuous quantitative differences among individuals were onmipresent in natural populations. At least at the phenotype level the minor variations that Darwin recognized as being the material on which natural selection operated were found to exist for essentially every character that was carefully measured. Undoubtedly to some extent this variability is caused by differential environmental influences operating during individual development. However, the existence of phenotypic individual differences suggested the possibility of equally widespread genotypic differences among individuals.

Supporting evidence came from the results of experiments that intentionally directed genetic selection or simply changed the breeding structure of existing populations without any intentional selection. In essentially all instances of these experiments, one or the other procedure resulted in a change over generations in the degree of expression of several characteristics. Such experimental results made sense if the existence of underlying genetic variation was assumed with relevance to most, if not all, traits.

Controversy continued between proponents of these two views until recently. Within the last decade methodological innovations have allowed for the first time direct assessment of the degree of genetic variation in existing natural populations. To date, the empirical results from species as diverse as fruit flies, mice, and men are congruent in overwhelmingly rejecting the classical view. Allelic variation at many loci appears to be the rule rather than the exception. Single "wild-type" alleles simply do not exist for many loci. Often no single gene can be reasonably characterized as the "normal species type." Instead, populations consist of sets of multiple alleles at many loci (Dobzhansky, 1970; Gooch & Schopf, 1973; Selander & Kaufman, 1973).

Any description of the historical course of evolution, therefore, should take into account the maintenance of genetic variation in natural populations, and experimental approaches to evolutionary questions are weakened to the extent that they fail to recognize the variability existing in natural populations, whether extant or extinct.

BEHAVIOR IN EVOLUTION

Against this background of omnipresent genetic variation must be placed another conceptual revolution that has taken place (or is underway). This one is directed to the roles of behavior, morphology, and genetics in evolution.

Darwin juxtaposed behavioral and morphological examples throughout his writings to illustrate the sufficiency of natural selection for evolution. In doing so he even avoided dualistic traps of the mind–body variety. However, after such auspicious beginnings considerations of the role of behavior in evolution were for a time almost nonexistent. Hard structures, of course, tend to fossilize a bit more frequently than do direct records of behavior, and reliable anatomical measurements can usually be made much more easily than can equally reliable measurements of behavioral characteristics. It is not surprising that the two main approaches to the study of evolution, the comparative–phylogenetic and the experimental–genetic, tended to concentrate early on anatomical and morphological traits instead of on behavioral ones.

This tendency to disregard behavior was reinforced for a time by the bias toward extreme environmental determinism that characterized the mainstream of psychology during much of this century. After all, if in a particular discipline genetic influences are claimed to be negligible by most of the specialists, it becomes quite reasonable for evolutionists seeking material to study to look elsewhere.

Since the publication of landmark volumes in the last decade or two (Fuller & Thompson, 1960; Roe & Simpson, 1958) an explosion of information has occurred at least with regard to genetics and behavior. Recent reviews and texts, and even the appearance of a topical journal, adequately document the assertion that with regard to genetic influence there is no known difference in principle between morphological and behavioral variables (DeFries, Ehrman, Gottesman, Schull, & Vandenberg, 1972; McClearn, 1970; McClearn & DeFries, 1973). Aspects of behavior by whatever classification, whether sensory or cognitive abilities, personality, temperament, social behavior, learning, and intellectual performance, among humans or other animals, display genetic involvement just as do anatomical, biochemical, or other phenotypical traits. Whether, beginning with morphologically identified single genes, pleiotropic effects on behavior are sought (Thiessen, Owen, & Whitsett, 1970) or, beginning with some behavioral activity, genetic involvement in the determination of individual performance is sought (Whitney, 1973; Whitney, McClearn, & DeFries, 1970), the outcome has been so consistent that it is no longer a source of rational contention. As one review recently put it: "The early mission of behavioral geneticists centered about the demonstration that genetic variation was an important determinant of most (all?) domains of behavior and, in large part, this goal seems to have been accomplished" (Lindzey, Loehlin, Manosevitz, & Thiessen, 1971, p. 79).

Simultaneously with the emergence of experimental behavior genetics, a shift has occurred in conceptualizations of the role of behavior in evolution. Instead of being largely ignored, or being viewed as a trivial set of characteristics that somehow have just tagged along as epiphenomena dependent on somehow more basic morphological change, behavior is coming to be viewed as an important facet in evolution, even in the evolution of nonbehavioral and nonnervous system characteristics. As Mayr (1963) points out:

> A shift into a new niche or adaptive zone is, almost without exception, initiated by a change in behavior. The other adaptations to the new niche, particularly the structural ones, are acquired secondarily (Mayr, 1958, 1960). With habitat and food selection—behavioral phenomenon—playing a major role in the shift into new adaptive zones, the importance of behavior in initiating new evolutionary events is self-evident. Sibling species, in spite of their morphological similarity, often show remarkable behavioral differences. Most recent shifts into new ecological niches are, at first, unaccompanied by structural modifications (Robson & Richards, 1936). Where a new habit develops, structural reinforcements follow sooner or later [p. 604].

In other words, behavioral modification may often precede evolutionary changes, at least in nonnervous system morphology.

With regard to the evolution of the nervous system and behavior it is especially important, indeed absolutely necessary, to avoid the traps of mind—body dualism. Because behavior is in the main the result of nervous system function, behavioral variants and nervous system variants are often rightly regarded as different manifestations of a single phenomenon. In the terminology of genetics, a particular behavioral phenotype and a particular nervous system phenotype, are often, if not usually, the pleiotropic manifestation of a single genetic constitution. As such, neither the morphological nervous system specialist nor the functional behavior specialist has an advantage over the other in the ease with which his or her material yields answers to evolutionary questions. Instead, the choice of which material to use is mostly a matter of personal interest. This realization is, of course, the main justification for the present attempt to integrate behavioral and neurological approaches to evolutionary questions and, conversely, to make use of evolutionary or genetic considerations in the study of behavior and neurology.

In this vein it should be immediately obvious that in the reconstruction of phylogenies neither behavioral nor morphological approaches have a general advantage; they are, or should be, isomorphic. As an illustration, some of the considerations underlying the recognition of homologies can be used. As Hodos has suggested in this volume, if in present representatives of two lineages phenotypically different behavioral activities strongly depend on the function of the same neural structure, then the activities may be thought of as homologous to each other. Conversely, if two phenotypically similar behavioral activities are mediated by different neural structures in two lineages, this is evidence that the activities are not homologous. Although the first statement may often be useful and indeed even true, the latter may not. Consider an initial population with variation in both structure and function; to some extent engaging in Behavior *A* is selectively advantageous, and the incidence of Behavior *A* is a function of variation in both Structure *X* and Structure *Y*. The initial population then becomes divided into two separate breeding populations, and Behavior A remains selectively advantageous in each population. However, the composition of the two subpopulations and the selective pressure on them is unlikely to be exactly the same and, therefore, the course of evolution cannot be exactly the same. In one population favorable variants in *X* may be more successful than favorable variants in *Y*, and vice versa for the second population. Selection may operate to maintain or enhance the "identical" Behavior A through accentuation of the *X–A* pathway in the first population and accentuation of the *Y–A* pathway in the second population. An eon or two after the initial separation of the two populations an observer encounters two separate descendant groups, both of which engage in a similar behavioral activity; however, the behavior is now mediated by different structures in the two groups, even though the behavior itself is, at least by most theoretical definitions, a perfect instance of homology. Obviously, neither a strictly behavioral nor a strictly morphological approach is advantageous when evolutionary questions are attacked.

The chief practical result of the foregoing suggestion that psychology is not different in principle from morphology when genetic involvement or evolution is considered is that the entire literature of population genetics and evolutionary biology is directly relevant to questions concerning the evolution of the nervous system and behavior. The remainder of this chapter serves only to present some particular highlights, which I have found particularly interesting or illuminating.

QUANTITATIVE POPULATION MODEL

To consider in overview how some genetic population processes operate in evolution and then some implications for studies of behavioral and neurological evolution, it is convenient to use the model of quantitative or biometrical genetics, which goes a bit beyond Mendelian genetics. An advantage of this approach is that the model is sufficiently general to apply to any number of genes that might affect a particular phenotypic trait, whether morphological or behavioral. Because the model is not completely obvious in all its aspects, a word regarding this approach is warranted.

Classical Mendelian models are useful for many theoretical derivations but are empirically testable only when the individual effects of single genes can be detected against the background of phenotypic variation caused by other genes and environmental effects. For many traits, differences between individuals are often quantitative rather than qualitative and simple patterns of single-gene Mendelian inheritance are often obscured from view. Quantitative genetics represents a statistical approach founded in basic Mendelian genetics, but its application is not dependent on identifying single-gene effects or prior knowledge of the number of genes that affect the trait of interest. When detectable single-gene effects are present, investigation of the effects can be profitably encompassed within the quantitative framework. However, the only prerequisite necessary for the application of quantitative genetic methods is the ability to reliably measure the phenotypes of interest. Obviously if a characteristic for an individual cannot be reliably measured, the inheritance of that characteristic among the individuals of a population cannot be traced. The following development of some basic quantitative genetic concepts relies heavily on the more complete descriptions by Falconer (1960) and by McClearn and DeFries (1973). Either of those sources and the references they contain can be consulted for derivations of the concepts and relationships that are of necessity only briefly described here.

The terminology of the basic quantitative genetic model begins with P, the phenotypic value of an individual for some particular characteristic. It represents an actual measurement of some trait of an individual: height, weight, IQ score, number of errors in a maze, volume of the pontine nuclei, or any characteristic that can be reliably measured. The essence of the model is that P is a function of G and E, where G represents a genotypic contribution and E an environmental contribution to the phenotypic value of an individual. It is important to realize that, for the purposes of the basic model, E is defined as being independent of G. In other words, G encompasses all effects on the phenotype traceable to the genotype, regardless of how the effect is mediated, whereas E represents all other effects on the phenotype. [In various natural situations interaction between particular genotypes and particular environments might often be of importance, and correlations may exist within a population between genotypes and aspects of the environments in which they exist. To be sure, $G \times E$ interactions and correlations can be accomodated in particular instances; however, they need not enter into the present model (Roberts, 1967a).]

If G and E are the only sources of influence on P, then the simplest and generally most useful model is the so-called "additive" one, in which for an individual $P = G + E$. As this is the description that is begun with for an individual, the phenotypic mean P of a particular population consisting of any number of individuals can be described in terms of the means of the same causal components. $\bar{P} = \bar{G} + \bar{E}$. Furthermore, the variance of the population is in general composed of the same components: $V_P = V_G + V_E + 2 \text{ cov } GE$. Here V represents the variance contributed by each source and cov is the covariance. Because for the purposes of illustration G and E have been assumed to be independent, there is no covariance between genotype and environment; therefore, for a particular trait in a particular population, $V_P = V_G + V_E$.

For evolutionary concerns it is the genetic component of the phenotypic value that is of most importance so that is where I shall concentrate my discussion. The genotypic value, G, refers to the contribution of an individual's genotype to its phenotype. However, among sexually reproducing diploid animals an individual does not transmit an intact genotype to its offspring. What is transmitted across generations is in most cases a virtually random sample of one–half of the genes that constitute the individual's genotype. The other half of an offspring's genes are contributed by the other parent. Because of the inheritance of samplings of single genes rather than diploid genotype sets, some value must be considered, at least in theory, the referent of which is a single gene rather than a genotype. This value is the average effect of a gene, and for an allelic pair of genes it is defined as the average phenotypic value of those individuals that received a particular allele from one parent minus the mean value of the entire population. The additive value of an individual, A, is the sum of all the average effects of the individual's single genes irrespective of the particular pairs in which they happen to occur; A represents part of G, the genotypic value of the individual, but is not necessarily all of G. The genotypic value of an individual is determined by the average effects of its genes and also depends on the interactions between the particular genes that constitute its unique diploid genotype. Interactions between genes at a locus is what is referred to as "dominance" in Mendelian terminology. If there is any dominance between the two alleles at a particular locus, the dominance affects some individual's genotypic value. However, the dominance effects are not simply transmitted to any offspring, because they result from the interaction of the two genes at a locus and each offspring receives only one of the alternative genes from a single parent.

For the contribution of one locus the genotypic value of an individual has two components. One of these components, A, is primarily responsible for the genetic resemblance between related individuals and the other component, D, results from the interactions between the specific alleles of the gene pairs that constitute an individual's diploid genotype. Obviously D, although a component of G, does not contribute in a simple way to the resemblance between a single parent and offspring, for genes, not gene pairs, are transmitted across generations. When more than one locus is considered, more terms should be included to accomodate interactions between alleles at different loci—that is, epistatic effects. In reality the various epistatic interactions are potentially many and complex, and although they can be handled both theoretically and experimentally, they are in practice most usually combined with the dominance term, yielding a partition of G into two components, A and *Not-A*, where A is the additive contribution to the genotypic value and the symbol D is retained for all the nonadditive interactions. It should be noted that because of the way they are defined and derived A and D are independent (that is, their covariance is zero) so the description of G that is sufficient for my purposes here is $G = A + D$. Because $P = G + E$ the relationship can now be expressed as $P = A + D + E$ and for the total observed phenotypic variance of the population, $V_P = V_A + V_D + V_E$ . . . plus appropriate covariance terms, if in a particular situation there is a correlation between particular genotypes and the environments in which they occur, and contributions to the total variance caused by nonadditive interactions between the various main genetic components and the environment, if such exist. The covariance and interaction terms can still be profitably ignored because what is sought is a partitioning of the observed phenotypic variance into the additive genetic variance versus all the other genetic and nongenetic sources of variation.

It is V_A, the additive genetic variance, that is of most interest because it is the variance of the additive genetic values, which are the major genetic cause of resemblance between relatives. It is also the genetic component of variation that is most directly involved in any response to selection. This partitioning of the total variance of a population into the additive genetic component versus all the rest, when expressed as the ratio of the additive variance to the total phenotypic variance, is called the "heritability" (h^2) of the trait under consideration: $h^2 = V_A/V_P$. Another quantity, representing the proportion of the phenotypic variance attributable to all genetic sources (V_G/V_P) is sometimes referred to as heritability "in the broad sense," or as the "coefficient of genetic determination." It is quite different from, and should not be confused with, h^2, heritability "in the narrow sense" of V_A/V_P. Finally, it should be noted that every value so far considered is strictly a population parameter; that is, for a given phenotypic characteristic everything including the average effect of a particular gene is specifically dependent on the frequency of all the other genes, the breeding pattern, and the environment of the population.

Because V_A is the variance of the additive values of the individuals that comprise a particular population, h^2 provides an index of how reliable the phenotype is as an indicant of the breeding value. As a proportion, h^2 can vary from 0 to 1, which indicates how well the deviation of an individual's offspring from the phenotypic mean of the entire population can be predicted from knowledge of the individual's phenotype. If, in an extreme case, $h^2 = 1.0$, then all of the phenotypic variance of the population for the trait under consideration results from the average effects of the genes in the population, $V_A = V_P$ for the population as a whole, and $P = A$ for an individual. When the individual phenotype is known exactly what the offspring is like is known. However, if $h^2 = 0$, then knowledge of an individual's phenotype provides no clue as to the average phenotype of its offspring; the best guess for the offspring is to simply use the average value of the entire population.

SELECTION

Two forms of selection that have been relatively well studied and that are considered in turn here are directional and stabilizing selection. Directional selection denotes the situation in which individuals at one end of a phenotypic distribution are more fit than the average for their population. "Fitness," of course, refers to Darwinian fitness (synonymous terms are adaptive value and selective value), which is the proportionate contribution of offspring to succeeding generations.

Selection, whether natural or artificial, is based on phenotypes. The selecting agent, whether aspects of the natural environment or man in the case of artificial selection, sees only the available phenotypes. To the extent that the individual phenotype provides an indication of how the individual's offspring differ from the average of the population, there can be a response to directional selection. Therefore, in addition to selection pressure a crucial variable is, as seen above, the heritability of the trait ($h^2 = V_A/V_P$). The additive value, in turn, is dependent on the average effects of the genes in the population, which is a function of all the gene frequencies in the population. In other words, the response of the population to selection (R), which is the change in the mean value of the population from one generation to the next, depends on the intensity of selection and on the heritability of the phenotypic trait on which the selection is based. The intensity of selection can be expressed as the selection differential (S), which is the difference between the average of the entire population and the average of those individuals that are the parents for the next generation. This relationship can be expressed symbolically as $R = h^2 S$, from which it can be seen that one of the ways by which the heritability of a trait can be estimated is from the results of a directional selection experiment: $h^2 = R/S$. The value estimated in this fashion is referred to as "realized heritability" and, for many people, constitutes the "proof of the pudding" for the heritability of any trait.

It must be remembered that the genetic values are population parameters dependent on the composition of the gene pool of the population. With each generation that responds to directional selection there should be a change in the frequencies of the genes in the population and V_A may therefore change. When directional selection is continued over generations in experimental situations there is often a gradual decrease in the response of the population from one generation to the next until, eventually, little or no further change occurs even though the selection pressure is maintained (Lerner, 1954).

Theoretically, a population can demonstrate such a "plateau," or selection limit, in the presence of a selection differential when V_A, and therefore h^2, approaches zero ($h^2 = R/S$; if $R = 0$ with $S > 0$, then $h^2 = 0$); this does not require an absence of genetic variation (V_G) affecting the phenotype, but only an absence of V_A. In other words, the population as a whole becomes equilibrated and no longer responds to directional selection with a change in mean value when the gene frequencies over all loci reach values such that they no longer contribute an additive component of variation. Much genetic variation may remain and be visible as genetic influences on individual phenotypes within the population but it is caused by dominance interactions and higher order epistatic interactions among the genes in the still heterogeneous gene pool of the population. From the theoretical account it is obvious that adaptation is a property of interbreeding populations rather than strictly a property of individuals.

During directional selection the genetic control of variability can, in effect, shift from being related to average effects of genes and become more related to interactive effects (dominance, etc.) of the genes in the gene pool. The occurrence of this phenomenon as a consequence of directional selection leads to the prediction that traits that have been directional components of Darwinian fitness in the recent evolutionary history of a population should have low heritabilities. This prediction is a correlary of Fisher's "fundamental theorem of natural selection," which states that the "rate of increase in fitness of any organism at any time is equal to its (additive) genetic variance in fitness at that time" (Fisher, 1958, p. 37). In a stable population (one that is adapted to its present environment and not changing over generations) the additive genetic variance of fitness itself is zero; therefore, the realized h^2 of fitness itself is zero in a stable population. Traits that are themselves directional components of fitness can therefore be expected to have little additive genetic variance (McClearn & DeFries, 1973; Roberts, 1967b).

These considerations are generally borne out by empirical data in that h^2 tends to be low for traits that are obvious directional components of fitness and substantially higher for traits less closely related to fitness. For instance, h^2 = .05 for conception rate in cattle and .85 for slaughter weight. Ear length in corn has an $h^2 \cong .17$, whereas for height of plants the $h^2 \cong .7$. Egg production in poultry and *Drosophila* has an $h^2 \cong .2$, whereas for litter size in mice the $h^2 \cong$ .15 (Dobzhansky, 1970).

Because the phenotypic expression of traits directionally related to fitness is largely mediated by interactions among the coadapted genes of the population gene pool, a procedure that reduces the opportunity for genic interactions to occur should, in theory at least, differentially affect components of fitness. Inbreeding is such a procedure. An increase in intermating among close relatives from a population (full siblings, mother–son, etc.) to form many different inbred families does not, in the absence of selection, change the frequencies of the genes from what they are in the base population. However, the distribution of genes among diploid genotypes does change. Within each inbred family line the homozygosity averaged across all loci increases. One line homozygously fixes some particular alleles, whereas another line at random fixes other alleles, and so on. Across many inbred lines all of the individual genes from the base population may still be represented, but they now exist as more homozygous sets, with necessarily less heterozygosity. Dominance effects can be expressed only when two different genes at a locus exist together in one individual; in other words, dominance can be expressed only in heterozygotes. Because of the overall loss of heterozygosity with inbreeding a progressive reduction occurs in opportunities for dominance and other interactions among the genes of the population. Conversely, the simple process of intercrossing already existing inbred lines does not change the frequency of genes in a total population consisting of all the inbred lines crossed, but it does change the distribution of the genes among diploid individuals. Inbred lines consist of individuals that are uniformly homozygous. The hybrids resulting from the crossing of two inbred lines are heterozygous for all loci at which the inbred parents are different. Crossing of inbred lines increases the degree of heterozygosity and therefore increases the opportunities for interactions among genes. Empirically, inbreeding depression and its obverse, heterosis or "hybrid vigor," is among the best established phenomenon in the basic biology of normally outbreeding organisms. The very terminology of inbreeding "depression" reflects the reduction with inbreeding of the phenotypic expression of metric characters that are directionally related to fitness. Similarly the "vigor" of hybrids is a statement reflecting the frequent increase in expression of fitness-relevant characteristics that follows the simple process of intermating individuals from different inbred lines. Even the ancients have been aware of the inbreeding depression that is now understood to result from disruption of a population's coadapted gene pool. Therefore, as is reduction in heritability, inbreeding depression and heterosis are correlaries of Fisher's fundamental theorem of natural selection.

Stabilizing Selection

Traits that are important components of fitness are not necessarily linearly related to fitness. For some characteristics an intermediate level of expression is optimal. For instance, among many animals it is selectively advantageous to be aggressive but not too aggressive. Among male mice (*Mus musculus*) individuals must be aggressive enough to establish dominance over other males if they are to reproduce (discussed below); however, if they are to reproduce they must not be so aggressive that they attack a cat. In natural situations differential fitness is the net result of many different, and often conflicting, selection pressures. Phenotypic characteristics that contribute to fitness but are not directional components are to some degree under the influence of stabilizing selection. In instances of artificial selection, say for increased egg production within a flock of poultry, the population may cease to respond with an increase in egg production when that trait can increase only at the expense of other traits that are also components of fitness. Because of the total physiology of individual birds, which is a function of the gene pool of the population, egg production cannot increase beyond a certain point because the best egg producers become less fit for other reasons than are individuals that produce fewer eggs. In other words, the directional selection continues to influence a trait until the directional pressure is balanced by counteracting aspects of selection. This is to say that a trait that is positively related to fitness initially may cease to respond to directional selection when the gene pool has been rearranged so that for further increases the trait becomes negatively correlated with fitness. The artificial selection continues to weed out the lowest egg producers whereas natural selection weeds out the highest egg producers and the flock as a whole does not change over generations. It should be remembered that $R = h^2 S$. With counteracting selection pressures the net S may be zero; therefore, R may be zero even though h^2 is substantial.

Usually, and contributing to the unhappiness of animal husbandmen, a relaxation of artificial selection results in a quick regression of the previously selected population to a lower level of performance than that previously maintained by artificial selection. This populational rebound results from the operation of natural selection that is no longer balanced by the artificial selection (Lerner, 1954). In natural populations a similar balance between conflicting selection pressures is often operative, resulting in relative stabilization of the population as a whole with regard to particular traits. Populations then remain stable or change slowly over generations for a particular trait not because there is little or no genetic variation affecting the phenotypic expression of the trait, but because the conflicting selection pressures on the total population balance out as stabilizing selection (Ford, 1971; Mayr, 1963). Such traits are not at the present time linearly (directionally) related to fitness and may display substantial additive genetic variance.

It is probably true that these processes rarely if ever go to completion in natural populations because of fluctuations in both genetic and environmental factors. Equilibration of local population gene pools can be constantly challenged by recurrent mutuations and gene flow from other semiisolated populations experiencing somewhat different selection pressures. Additionally, selection differentials may often vary as a function of such factors as gene frequency, population density, both temporal and spatial habitat differences, and stage of individual development. All of these fascinating subjects are currently the topics of much active research and theorizing. It appears that they generally operate to further maintain and enhance the genetic diversity of the gene pools of populations.

The complexities are many and consideration of genetic processes in evolution has not congealed into a uniformly accepted dogma. At the same time, generalizations of the sort offered above are founded in much theory with a plethora of supportive empirical evidence. Applied with appropriate caution, information concerning genetic processes in evolution can be of use to those interested in the phylogeny of particular systems.

One very straightforward application is to utilize information concerning the genetic architecture underlying traits of extant species together with information about current selection pressures to make inferences concerning the role of the character in the recent evolutionary history of a lineage. Another straightforward application is to take the general genetic information into consideration when designing research strategies for asking evolutionary questions that are not themselves explicitly genetic. An example of each application is given in turn.

Intelligence

The phenotype of human intelligence, at least as measured by existing intelligence tests, can be used to illustrate the application of genetic information to evolutionary interpretation. Intelligence, defined as a metric phenotype either as a global IQ score or as scores on a set of subscales, is well suited for this purpose because of the statistical reliability of the available tests. So-called "intelligence" tests are by no means perfectly reliable; however, as it turns out most modern ones are more reliable than any other indices of complex human behavioral functioning yet in existence. Reliability of individual measurement is, as pointed out above, a basic requirement of genetic analysis.

Another advantage of using intelligence test scores as an example is the sheer mass of data that exists from many studies. For instance, one very short review surveyed the results of 52 studies that contained an aggregate of over 30,000 correlational pairings after investigations that were inadequate in design or too specialized in subjects sampled had been excluded (Erlenmeyer-Kimling & Jarvik, 1963).

One result of this mass of research is the conclusion that the heritability of intelligence in populations of present-day industrialized societies is definitely nonzero and is probably quite high. Estimates of the coefficient of genetic determination, or heritability in the broad sense of V_G / V_P, vary from about .8 (Jensen, 1969) to about .5 (McClearn & DeFries, 1973). For any single population, it has been impossible to arrive at precise numerical estimates of h^2 that are acceptable to all investigators because of the existence in the natural environment of genotype-environment correlations and Genotype X Environment interactions that may influence the development of the trait. In any event, virtually all serious investigators agree that much of the genetic variation appears to be additive so the narrow sense heritability (V_A / V_P) approaches the broad sense heritability and may be .5 or greater (Vandenberg, 1971).

These high estimates of h^2, combined with an application of the rule of thumb (from Fisher's fundamental theorem) that traits which are linear components of fitness tend to have low heritabilities, have led to suggestions that, perhaps surprisingly, whatever is measured by current IQ tests has not been an important directional component of fitness in the recent evolutionary history of man (McClearn & DeFries, 1973).

By this suggestion, whatever may be the functional characteristics related to the explosive increase in brain size in the recent lineage of man, they are not what is being measured by present IQ scores.

Furthermore, a quite consistent pattern of h^2 differences across subscales of intelligence tests has also been found (Vandenberg, 1968). Scales labeled "numerical," "spatial," and "verbal ability" and "word fluency" tend to yield high heritability estimates, whereas reasoning and memory factors are much lower and often negligibly heritable. From this patterning it has been suggested (McClearn, 1968; McClearn & DeFries, 1973) that perhaps the memory and reasoning scales are measuring functional abilities that have been closely related to fitness whereas the other scales are not. The overall IQ score and subscales that yield high h^2 values may represent phenotypes reflecting characteristics that have been under stabilizing selection pressure for an intermediate optimum with selection against both extremes would be compatible with the maintenance of a high heritability coefficient (McClearn, 1968).

The most serious criticisms of these seemingly straightforward inferences from the magnitude of the h^2 estimates stem from the nature of the populations under consideration. As emphasized above, the value of h^2 of any trait is not a constant; it is a population parameter (as is a mean or a variance, etc.) and therefore may be expected to change with a change in the population. However, it should be noted that the lability of h^2 values for polygenically influenced characteristics may not be as great as is sometimes suggested. Empirical evidence from both behavioral and morphological phenotypes indicates that h^2 estimates may often be very consistent for similar phenotypes in different populations within a species and even across related species (Falconer, 1960; Whitney *et al.*, 1970).

Be this as it may, direct inference of past selective value based on the magnitude of present heritability assumes that the current population is not appreciably different, in genetic and environmental aspects that could effect IQ scores, from what it was in the recent phylogenetic past. Changes in some characteristics of the population may result in changes in the heritability of a trait in directions that are broadly predictable. Each of a number of quite gross factors that are known to be operative in recent human populations can tend to increase the h^2 of a trait that has been a major directional component of fitness.

If important aspects of the environment are changing over generations, then the heritability of a trait that is a directional component of fitness can be quite substantial. The h^2 of the fitness components would be quite high as the adaptations of the gene pool tracked the environmental changes. There is no question but that many aspects of the environment in which modern man exists have changed over the last few thousand years. The directional importance of the changes with regard to levels of IQ scores are, however, open to question. In addition to possible cumulative directional changes in relevant environmental factors, a general relaxation of many selection pressures such that the overall fitness of a population increased could lead to an increase of heritability of traits that had become, under more stringent circumstances, much less heritable.

The tremendous increase in numbers of the human population in historical times is ample evidence that the overall fitness of human populations has, by definition, increased. In such an unstable population experiencing a great increase in numbers the additive genetic variance of fitness itself, and therefore of fitness-related traits, can be quite substantial. It is interesting to note that in several species important microevolutionary shifts in previously stable characteristics have been observed to occur during the reduction in numbers following even short episodes of population increase (Ford, 1971). Similarly, in artificial selection one of the tricks employed to break out of a selection limit is to relax the selection pressure for a few generations, thus allowing an increase in population size to occur. Genetic recombination during the generations of relaxed selection is allowed to occur so that when selection pressure is reinstituted the population may again respond with gains in the selected traits.

Quite apart from the aspects of population composition that relate directly to environmental matters, more strictly genetic factors must be considered. If the amount of gene flow among previously semi-isolated local populations were to increase, the newly constituted population resulting from this increased genetic migration could quite literally begin an entirely new evolutionary direction. This phenomenon can occur without any concomitant change in environmental pressures other than those that have allowed the migration to occur.

There is little question that in recent centuries the genetic admixture has greatly increased for many human populations. The concept of America as a very recent "melting pot" of individuals from many previously diverse and partially isolated populations is to some extent accurate at the genetic level. Even within local regions of Europe for which data exist, the mating patterns of inhabitants have shifted toward increased gene flow with the increased ease of contact among residents of separate localities (Cavalli-Sforza & Bodmer, 1971). When a species is subdivided into a number of partially isolated breeding populations the genetic effects of selection must be considered separately for each of the isolates. For a polygenically influenced trait, such as an intelligence test score, the gene frequencies across many different loci could be quite different in different isolates even if each population was near a directional selection limit. The average phenotypic level of each isolate for some trait could conceivably be quite similar, and the heritability of a trait could be very low within each isolate; at the same time, however, the particular configuration of the gene pool could vary from one subpopulation to another. The creation of a new population by merging subpopulations that had previously been somewhat isolated could result in a substantial heritability for a trait that was a directional component of fitness and had had a low heritability in each of the earlier subpopulations.

A directly analogous procedure is a well-known trick employed by animal husbandmen in artificial selection programs. Selection is performed within a closed population for a number of generations until a diminution of the response to selection occurs. Gene flow is then instituted by breeding members of the selected population to individuals from another population that is undergoing concomitant selection, or even to individuals from an unselected population. Following this influx of genes from another group a further response to selection is often obtained. That this further response occurs without a change in the artificial selection pressure indicates that the influx of new genes has resulted in an increase in the h^2 of the trait. To what extent genetic admixture of previously semi-isolated populations contributes to the high heritability of an intelligence test score among some present human populations is unknown. However, the historical record of population change is consistent with this possibility and the notion is to some extent testable as long as relatively isolated groups with substantial historical continuity exist.

In the main it must be concluded that inferences of evolutionary relevance from heritability alone are a one-way game. If a phenotype when measured reliably displays a low heritability, the inference that it has in recent evolutionary time been a major directional component of fitness appears justified. However, if the phenotype displays a high heritability in existing populations, that by itself is not sufficient to conclude that the trait is unrelated to fitness.

Additional information about the genetic architecture of the trait that is relevant to fitness interpretation comes from the effects of inbreeding and hybridization. Direct investigation of changes in the average value of the trait that occur as a concomitant of particular patterns of mating among members of a population should yield meaningful information because inbreeding depression and its converse, heterosis, are often observed for metric traits that are components of fitness. Quite a lot of information exists concerning the effects of inbreeding on intelligence. That consanguineous matings result in an increased incidence of mental deficiency is a well-known phenomenon that is part of the folklore of most primitive societies. Indeed, the existence of incest taboos in virtually all human societies may represent a cultural response that has originated in part to minimize the impact of inbreeding depression on intelligence and other phenotypic traits (Lindzey, 1967). McClearn and DeFries (1973) and Vandenberg (1971) summarize information from several independent sources that is congruent in suggesting that inbreeding leads to lower IQ scores. For instance, the prevalence of mental retardation among offspring of cousin marriages is over 3.5 times greater than among offspring of unrelated controls (Böök, 1957).

In an investigation involving over 1800 children of parents who practiced inbreeding to various degrees, Schull and Neel (1965) found that the average scores on all of the 13 subscales of an intelligence test were depressed by inbreeding. Although some subtests were more affected than others, a 10% increase in inbreeding corresponded with a 7% reduction in mean IQ. For example, on the average the offspring of brother–sister matings would be expected to score about 18% less on IQ than children of unrelated parents (McClearn & DeFries, 1973). Such data, indicating that intelligence is susceptible to substantial inbreeding depression, clearly suggests that intelligence has to some extent been of positive selective value in the recent evolutionary history of man.

Another source of genetic information relevant to evolutionary interpretation comes from investigating the relationship of intelligence to fitness in present populations. Among industrialized societies a frequent observation is that parents of lower intelligence tend to produce larger families than do parents of higher intelligence. These observations are a major source of the recurrent concerns of lay eugenicists lamenting the possibly dysgenic effects of modern societal conditions on human evolution. At the same time, however, large-scale surveys conducted within this century have reported a small but very significant increase in intelligence test scores across consecutive generations (Benjamin, 1966; Carter, 1966). Although such increases may reflect the impact of changes in environmental conditions, such as increased communication and better schools that are slightly overcompensating for opposite genetic trends, the apparent paradox between large family–low IQ and slight increases in average IQ across the generations has been resolved independently by Higgins, Reed, and Reed (1962) and by Bajema (1963). These investigators took into account the reproductive performance of all members of their samples, not just those individuals who did have one or more children. There does appear to be a correlation between family size and intelligence with low-IQ parents tending to produce larger families, but at the same time there is a relationship between IQ and reproduction in that the proportion of individuals who have any children at all increases with increases in IQ. When individuals who do not reproduce are considered along with those who do have children, there is a clear positive relationship between differential fertility and intelligence. Even under conditions of modern society, therefore, there may be selection occurring for increases in intelligence (Eckland, 1972; Falconer, 1966; Neel & Schull, 1972). Neel (1970) has presented data suggesting that the positive selective advantage for intelligence may have been even greater in the past, for in some existing "primitive" societies village leaders disproportionately overreproduce compared to subordinate males, and low intelligence probably does not facilitate becoming or remaining a leader. Regardless, the bulk of the available genetic evidence does indicate that among humans intelligence has not only been a selective advantage in the recent evolutionary past, but perhaps surprisingly a measurable selective differential with regard to intelligence may be operative at present.

Domestication

The importance of taking genetic information into consideration in the design of research to answer evolutionary questions can be illustrated with regard to the products of domestication. It should be noted at the outset that this discussion is much more a presentation of problems than a catalog of solutions. It is a task of the first importance for present and future phylogenists to address themselves to solutions. The use of domestic animals as objects for evolutionary discourse has a long history, going back at least as far as Darwin. However, the present sorts of considerations may become increasingly important in the near future, especially with regard to studies of the evolution of behavior and nervous system structures, for the practice of using captive-bred subjects in experimentation seems bound to increase as nonhuman primates and many other animals come closer to extinction in their dwindling natural habitats.

Two aspects of the use of captive-bred animals in evolutionary research must be carefully separated. The first is the study of changes in populations under domestication as exemplars of evolution in action. The second is the study of domestic animals to gain information concerning the course of evolution and the importance, or indeed the very existence, of particular adaptations in various phylogenetic lineages. It is the very success of the first of these endeavors that presents difficulties for the second.

The rapidity of evolutionary change of many characteristics in domestication is well known and involves not only "superficial" characteristics but also traits that are centrally important to the survival of the wild progenitors of domestic varieties. Behavioral traits have changed drastically in numerous domestic lineages to the point that the behavioral patterns of domestic varieties can be lethal in natural settings. Lorenz (1974) mentions that among several species of fishes that in nature practice complex brood care, the parental behaviors have been lost or seriously disrupted after only a few generations of breeding in captivity. Similarly, some quail (domestic varieties of *Coturnix coturnix*) and numerous strains of domestic chicken (*Gallus domesticus*) have changed such that they no longer brood their eggs—they have lost a behavioral trait that is crucial to fitness under natural conditions.

Nowhere is the difference in expectations between holders of a typological view and a populational view more apparent than in considerations of domestic animals. Everyone reading this volume and its companion is by now thoroughly aware of the fact that a domesticated goldfish (*Carassius auratus*) does not represent for all purposes a fish-type ancestor in any phylogenetic series, nor does a rat or a turtle or a pigeon represent a prototypical mammal, reptile, or bird. However, many investigators should like to maintain that a goldfish from some single domestic population is at least representative of some wild species of fish, that a laboratory rat is representative of wild *Rattus norvegicus,* that domesticated pigeons at least represent free-living *Columba livia,* and that a rhesus monkey from a primate breeding center can be a safe standin for a wild *Macaca mulatta.*

Unfortunately, such assumptions are precarious for any characteristics and may be especially dangerous with respect to behavioral traits and associated nervous system structures. A central problem is that evolutionary changes that occur as a result of captive breeding may be very rapid, which in turn suggests that existing selection pressures in nature may be very strong with regard to many aspects of behavior. Examples of very strong selection pressures involving behavioral traits in wild populations are proliferating as methods improve for the recognition of individuals in field studies. In a wild colony of elephant seals (*Mirounga angustirostris*) 4% of the males have been reported to inseminate 85% of the females (LeBoeuf & Peterson, 1969). Among some natural bird populations the single factor of social status has been reported to be an almost perfect predictor of survival and reproductive potential, even though the proximal causes of failing to reproduce and death are many (Wynne-Edwards, 1968). Similarly, for many primate species, including *Homo sapiens,* reproductive performance is definitely not random and is not independent of behavioral characteristics.

In view of the considerations discussed above that many species do in fact maintain much genetic diversity and that genetic variation is often manifest in phenotypic behavioral variation, and now that behavioral characteristics may often be centrally important in differential reproductive success, it is apparent that captive-bred individuals may represent a very aberrant sample with regard to any behavioral trait and associated nervous system structures.

To avoid quick and obvious change of characteristics in captive colonies, breeders nowadays often attempt to minimize the degree of inbreeding in their colonies, or attempt to explicitly "random breed" their stocks. However, many wild animal populations appear to be definitely not random breeding with regard to many behavioral characteristics. Because little is known about the determinants of reproductive success among natural populations of almost every species, there are bound to be differences in the selective value of various traits and differences in population structure between wild and captive groups. There is no escaping the conclusion that these differences may result in the divergence of wild and domesticated populations for many characteristics. Indeed, it has been suggested that wild and domesticated rats are so different in numerous behavioral and physiological characteristics that they may well be considered separate species if they were not known to interbreed (Thiessen, 1972).

Lorenz (1971) has suggested that there is some consistency in the direction of changes that occur during domestication. He points out that the most consistent change is an increase in the range of variation of many traits, as is to be expected from the relaxation of constant stabilizing selection pressures that are operative in wild populations. Even among reindeer (*Rangifer tarandus*) this effect of domestication on phenotypic variation is apparent "despite the fact that this animal lives in its original environment and in complete freedom. Its environmental conditions differ from those of the wild form in (not even thorough) protection from wolves and a certain selection of breeding stags; the Laps in fact castrate the strongest stags in order to reduce their animosity" (Lorenz, 1971, p. 232).

It is interesting to note that basic differences in performance between individuals from wild and domesticated populations may in large part be at the root of some of the differences in theoretical approach and experimental emphasis that have resulted in obfuscating polemics between the early ethologists and comparative psychologists. At the present time it is important to realize that the existence of some particular behavioral tendency among domesticated animals does not assure the existence of that same behavioral tendency among the individuals of any wild population of the same species, to say nothing of its state among the phylogenetic ancestors or descendants of that species. Conversely, if some domesticated animals do not display a behavioral pattern or appear incapable of learning under some set of conditions, that is not strong evidence that wild individuals of the same species have similar deficits. In the context of a review of well over 100 genetic experiments concerned with animal learning, all of which have utilized various domesticated genotypes, Wahlsten (1972) has addressed himself to the question of whether conspecific animals of different genotypes learn at different rates but in basically the same way, or whether their performance indicates that processes may be fundamentally different across several genotypes. His conclusion was that a general learning process was not demonstrated because there was little if any consistency of genotypes across various procedures. Genotypes that did very well in some conditions were found to be quite inferior in others. That degree of variation was within species, not across different species.

Attempts to experimentally investigate the phylogenetic adaptive function of a characteristic with domestic subjects are even more precarious than are attempts to answer normative questions with domesticated animals. For example, from work with domestic strains of the house mouse (*Mus musculus*) my associates and I have suggested that ultrasonic vocalizations are of adaptive value for wild mice during mating. In the presence of females many male domestic mice produce ultrasonic signals that are similar in frequency and intensity to the ultrasonic calls of preweanling pups. Because female mice have a postparturition estrous period during which they are often impregnated and because parous females display heightened aggression, we have suggested that the infantlike ultrasounds produced by males are the result of sexual selection and function during the early stage of courtship among wild mice to inhibit female aggression and withdrawal if not to elicit approach (Whitney, Coble, Stockton, & Tilson, 1973). Furthermore, the genetics of ultrasound production among adult domestic mice is consistent with the hypothesis that this characteristic has been of some selective advantage (Coble, 1972).

Under conditions of domestication that included an elimination of direct male competition for mating such a courtship display system might be nonfunctional, and its components could therefore drift or be selected to new levels for other reasons. The results of a direct attempt to experimentally investigate the adaptive function of this communicative system with domestic mice could be quite misleading. It is well documented that adult wild house mice normally can readily hear ultrasounds of the frequency and intensity produced by conspecifics (Heffner, Van Oeveren, & Masterton, 1971; Ralls, 1967). However, adults of some domesticated varieties may be incapable of hearing the ultrasounds produced by infants and courting males. One of the domestic varieties (the DBA) that has been reported to have a high-frequency hearing deficit (Ralls, 1967) readily produces ultrasounds under many test conditions (Whitney, Alpern, Dizinno, & Horowitz, 1974). It may be that this variety represents an instance in which components of both sender and recipient mechanisms that comprise a communicative system of adaptive value to wild animals have independently acquired altered levels of functioning during domestication. Experimental investigation with a limited number of domestic varieties would establish very little concerning the function of this behavioral system among wild animals of the same species.

At the same time, it is probably too extreme to suggest that all behavioral patterns of domestic animals have been grossly altered by domestication, and it is too early to predict what kinds of behavioral characteristics are most susceptible to modification under domestication. For instance, as a counter example it has been found that among long-domesticated inbred mouse strains male dominance order can account for over 90% of the variation in male reproductive success (DeFries & McClearn, 1970), even though adult male competition has not been involved in the mating success of these animals for literally hundreds of generations. Additionally, in this instance the results with domestic mice are congruent with findings from wild groups and are predicted on the basis of strictly genetic studies of wild populations (DeFries & McClearn, 1972; Selander & Yang, 1970).

Anyone operating under the influence of remnants of mind–body dualism may be inclined to hope that nervous system structures are somehow less susceptible to variation within and across different domesticated populations than are aspects of behavior. Such ideas may be especially appealing from a pragmatic point of view, for with present techniques the laboriousness of even relatively gross histological procedures tends to encourage small-sample research. Examples with regard to nervous system structures are harder to come by, perhaps because of the typological nature of much of this research, but evidence is accumulating that nervous system morphological phenotypes are not different in these respects from behavioral phenotypes. Illustrative material ranges from the wide variety of so-called "neurological mutants" that have been described among domestic animals (Sidman, Green, & Appel, 1965) to the ease of selective breeding for high and low brain weight within a single species (Collins, 1970; Fuller & Geils, 1972). Aspects of brain configuration, from the relative volume of different parts of the forebrain (Wimer, Wimer, & Roderick, 1969) to the very presence or absence of such major fiber tracts as the corpus callosum (Wahlsten, 1974), are known to vary among domesticated individuals of single species.

Functional studies of intraspecific nervous system variations that are grounded in genetics have barely begun but include Hegmann's work with speed of neural conduction (Hegmann, 1972; Hegmann, White, & Kater, 1973) and investigations related to the topography of visual pathways within a number of different mammalian species (Creel & Giolli, 1972; Guillery, 1971; Guillery & Kaas, 1971; Hegmann, Kieso, & Hartman, 1974; Lund, 1965).

As captive-bred individuals from primate and other presently "exotic" species become increasingly available for behavioral and neurological investigation, astute students of evolution are presented with a unique opportunity to empirically investigate both the phylogeny and current evolution of nervous system structures and behavior. To take full advantage of this new world of experimental material their research designs and interpretation of results of necessity must include dynamic population genetic considerations to an extent that has been all too rare in the past.

SUMMARY

To the evolutionist, whose habitual thought patterns are keyed to the broad sweep of phylogeny, the two main points of this chapter may at first consideration be disconcerting.

The first main point is the emphasis on genetic variation as an omnipresent factor in natural populations. The second main point is that selection operating on that variation can, and perhaps often does, result in evolutionary changes of a rapidity that is likely to have been scoffed at by most investigators just a few years ago.

About the first point little more need be said; James Murray (1972) excellently captured the spirit of current population genetic thought:

> A century after the publication of *On the Origin of Species,* Darwin's assessment of the extent of genetic variation in natural populations has been finally vindicated. First by means of indirect experiments and then by more direct demonstration of biochemical polymorphisms, it has become abundantly clear that the diversity within the gene pool of many species is very great.
>
> It is also evident that a large part of the genetic variation is influenced by natural selection [p. 111].

The second point may be more difficult for the phylogenist, who attempts to think in time spans of tens and hundreds of millions of years, who knows that some whole faunas appear from the paleontological record to have changed very little over many millions of years, and who knows of instances of extant species that appear identical to fossils that are hundreds of millions of years old. However, the phylogenist also knows of relatively short time spans during which faunal changes took place with dramatic rapidity.

What may be the single most important finding from recent evolutionary genetic research is that selection pressures in nature are often much stronger than even the extremists of a few years ago would have suggested.

Apparent stability of populations over time is usually and perhaps always the result of constant, *strong,* selection pressures, and strong selection operating on genetically variable populations can lead to rapid evolutionary changes: "In his early work, in order to prove that natural selection could produce evolution, Fisher assumed and accepted extraordinarily low values for differences in fitness. We now know that such low values are quite likely the exception and that we have extraordinarily strong differences in fitness under certain circumstances . . ." (Mayr, 1967, p. 47). The empirical evidence that inescapably leads to this conclusion includes, in addition to the results of domestication considered above, the finding that five different endemic species of fishes exist in one lake that has been geologically isolated for not more than 5000 years (Greenwood, 1965). Johnson and Selander (1964, 1971, 1973) have investigated populations of the English sparrow (*Passer domesticus*), many of which have arisen from recent introductions into new areas. Numerous populations now exist, less than 100 years after introduction, that are morphologically distinct enough to be considered good subspecies by every ornithologist, (Mayr, 1967). Among insects Bush (1969) has presented strong evidence for the existence of virtually instantaneous and sympatric speciation that is, by the way, behaviorally mediated in important respects. The similarly rapid evolution of industrial melanism and pesticide resistance in many populations is well known.

E. B. Ford, who devoted a lifetime to investigations of evolutionary genetics, mentioned in the conclusion of the third edition of his work *Ecological Genetics* (1971):

> One of the most far-reaching results of recent work on ecological genetics is the discovery that unexpectedly great selective forces are normally operating to maintain or to adjust the adaptations of organisms in natural conditions . . . Major Leonard Darwin told me of a conversation with his father, the great Charles Darwin, who expressed his belief that by choosing the right material it might be possible actually to detect evolutionary changes taking place in natural conditions at the present time. For this purpose he said that long-continued investigations and careful records would be needed, extending over a period which he estimated at perhaps fifty years in species reproducing annually. As usual, Darwin was right, but on this occasion he was too pessimistic. The techniques of ecological genetics have made it possible not only to observe evolutionary changes in three or four generations in such forms but to evaluate the force of selection in wild populations and to analyze its immediate effects [pp. 360–363].

Exciting possibilities for a new understanding of behavior and nervous system structures are opened up by the full realization that the dynamics of their evolution are available and susceptible to experimental investigation. At the same time, attempts to reconstruct phylogenies from investigation of only a small sample of extant forms may be perilous because any analysis of evolution that is inconsistent with genetics cannot possibly be right.

REFERENCES

Bajema, C. Estimation of the direction and intensity of natural selection in relation to human intelligence by means of the intrinsic rate of natural increase. *Eugenics Quarterly*, 1963, **10**, 175–187.

Benjamin, B. Social and economic differentials in fertility. In J. E. Meade & A. S. Parkes (Eds.), *Genetic and environmental factors in human ability.* New York: Plenum Press, 1966. Pp. 177–184.

Bőők, J. A. Genetical investigation in a north Swedish population: The offspring of first-cousin marriages. *Annals of Human Genetics,* 1957, **21**, 191–221.

Bush, G. L. Sympatric host race formation and speciation in frugivorous flies of the genus *Rhagoletis* (Diptera, Tephritidae). *Evolution,* 1969, **23**, 237–251.

Carter, C. O. Differential fertility by intelligence. In J. E. Meade & A. S. Parkes (Eds.), *Genetic and environmental factors in human ability*. New York: Plenum Press, 1966. Pp. 185–200.

Cavalli-Sforza, L. L., & Bodmer, W. F. *The genetics of human populations.* San Francisco: W. H. Freeman, 1971.

Coble, J. R. "The relative contributions of genotype and environment to the production of ultrasounds by adult *Mus musculus* in male–female pairings." Doctoral dissertation, Florida State University, Tallahassee, 1972.

Collins, R. A. Experimental modification of brain weight and behavior in mice: An enrichment study. *Developmental Psychobiology,* 1970, **3,** 145–155.

Creel, D. J., & Giolli, R. A. Retinogeniculo–striate projections in guinea pigs: Albino and pigmented strains compared. *Experimental Neurology,* 1972, **36,** 411–425.

DeFries, J. C., Ehrman, L., Gottesman, I. I., Schull, W. J., & Vandenberg, S. G. Introductory statement. *Behavior Genetics,* 1972, **2,** 1.

DeFries, J. C., & McClearn, G. E. Social dominance and Darwinian fitness in the laboratory mouse. *American Naturalist,* 1970, **104,** 408–411.

DeFries, J. C., & McClearn, G. E. Behavioral genetics and the fine structure of mouse populations: A study in microevolution. In T. Dobzhansky, M. K. Hecht, & W. C. Steere (Eds.), *Evolutionary biology,* Vol. 5. New York: Appleton-Century-Crofts, 1972. Pp. 279–291.

Dobzhansky, T. On types, genotypes and the genetic diversity in populations. In J. N. Spuhler (Ed.), *Genetic diversity and human behavior*. Chicago: Aldine, 1967. Pp. 1–18.

Dobzhansky, T. *Genetics of the evolutionary process.* New York: Columbia University Press, 1970.

Eckland, B. K. Evolutionary consequences of differential fertility and assortative mating in man. In T. Dobzhansky, M. K. Hecht, & W. C. Steere (Eds.), *Evolutionary biology*, Vol. 5. New York: Appleton-Century-Crofts, 1972. Pp. 293–305.

Erlenmeyer-Kimling, L., & Jarvik, L. F. Genetics and intelligence: A review. *Science,* 1963, **142,** 1477–1479.

Falconer, D. S. *Introduction to quantitative genetics.* New York: Ronald Press, 1960.

Falconer, D. S. Genetic consequences of selection pressure. In J. E. Meade & A. S. Parkes (Eds.), *Genetic and environmental factors in human ability.* New York: Plenum Press, 1966. Pp. 219–232.

Fisher, R. A. *The genetical theory of natural selection.* (2nd ed.) New York: Dover Publications, 1958.

Ford, E. B. *Ecological genetics* (3rd ed.). London: Chapman & Hall, 1971.

Fuller, J. L., & Geils, H. D. Brain growth in mice selected for high and low brain weight. *Developmental Psychobiology,* 1972, **5,** 307–318.

Fuller, J. L., & Thompson, W. R. *Behavior genetics.* New York: John Wiley & Sons, 1960.

Gooch, J. L., & Schopf, T. J. M. Genetic variability in the deep sea: Relation to environmental variability. *Evolution,* 1973, **26,** 545–552.

Greenwood, P. H. The cichlid fishes of Lake Nabugabo, Uganda. *Bulletin of the British Museum, Natural History,* 1965, **12,** 315–357.

Guillery, R. W. An abnormal retinogeniculate projection in the albino ferret. *Brain Research,* 1971, **33,** 482–485.

Guillery, R. W., & Kaas, J. H. A study of normal and congenitally abnormal retinogeniculate projections in cats. *Journal of Comparative Neurology,* 1971, **143,** 73–100.

Heffner, H., Van Oeveren, R., & Masterton, B. Hearing in two species of wild rodents. *Journal of the Acoustical Society of America,* 1971, **50,** 148.

Hegmann, J. P. Physiological function and behavioral genetics. I. Genetic variance for peripheral conduction velocity in mice. *Behavior Genetics,* 1972, **2,** 55–67.

Hegmann, J. P., Kieso, R. A., & Hartman, H. B. Gene differences influencing visual system function and behavior. *Behavior Genetics,* 1974, **4,** 165–170.

Hegmann, J. P., White, J. E., & Kater, S. B. Physiological function and behavioral genetics. II. Quantitative genetic analysis of conduction velocity of caudal nerves of the mouse, *Mus musculus. Behavior Genetics,* 1973, **3,** 121–131.

Higgins, J. V., Reed, E. W., & Reed, S. C. Intelligence and family size: a paradox resolved. *Eugenics Quarterly,* 1962, **9,** 84–90.

Jensen, A. R. How much can we boost IQ and scholastic achievement. *Harvard Educational Review,* 1969, **39,** 1–123.

Johnson, R. F., & Selander, R. K. House sparrows: Rapid evolution of races in North America. *Science,* 1964, **144,** 548–550.

Johnson, R. F., & Selander, R. K. Evolution in the house sparrow. II. Adaptive differentiation in North American populations. *Evolution,* 1971, **25,** 1–28.

Johnson, R. F., & Selander, R. K. Evolution in the house sparrow. III. Variation in size and sexual dimorphism in Europe and North and South America. *American Naturalist,* 1973, **107,** 373–390.

LeBoeuf, B. J., & Peterson, R. S. Social status and mating activity in elephant seals. *Science,* 1969, **163,** 91–93.

Lerner, I. M. *Genetic homeostasis.* London: Oliver & Boyd, 1954.

Lindzey, G. Some remarks concerning incest, the incest taboo, & psychoanalytic theory. *American Psychologist,* 1967, **22,** 1051–1059.

Lindzey, G., Loehlin, J., Manosevitz, M., & Thiessen, D. Behavioral genetics. *Annual Review of Psychology,* 1971, **22,** 39–94.

Lorenz, K. *Studies in animal and human behavior,* Vol. 2. Cambridge, Mass: Harvard University Press, 1971.

Lorenz, K. *Civilized man's eight deadly sins.* London: Methuen, 1974.

Lund, R. D. Uncrossed visual pathways of hooded and albino rats. *Science,* 1965, **149,** 1506–1507.

Mayr, E. Behavior and systematics. In A. Roe & G. G. Simpson, (Eds.), *Behavior and evolution.* New Haven: Yale University Press, 1958. Pp. 341–362.

Mayr, E. The emergence of evolutionary novelties. In S. Tax (Ed.), *The evolution of life.* Chicago: University of Chicago Press, 1960. Pp. 349–380.

Mayr, E. *Animal species and evolution.* Cambridge, Mass.: Belknap Press, 1963.

Mayr, E. Evolutionary challenges to the mathematical interpretation of evolution. In P. S. Moorhead & M. M. Kaplan (Eds.), *Mathematical challenges to the neo-Darwinian interpretation of evolution.* Philadelphia: Wistar Institute Press, 1967. Pp. 47–54.

McClearn, G. E. Social implications of behavioral genetics. In D. C. Glass (Ed.), *Genetics: Biology & behavior.* New York: Rockefeller University Press & Russell Sage Foundation, 1968. Pp. 164–168.

McClearn, G. E. Behavioral genetics. *Annual Review of Genetics,* 1970, **4,** 437–468.

McClearn, G. E., & DeFries, J. C. *Introduction to behavioral genetics.* San Francisco: W. H. Freeman, 1973.

Merrell, D. J. Limits on heterozygous advantage as an explanation of polymorphism. *Journal of Heredity,* 1969, **60,** 180–182.

Murray, J. *Genetic diversity and natural selection.* Edinburgh: Oliver & Boyd, 1972.

Neel, J. V. Lessons from a "primitive" people. *Science,* 1970, **170,** 815–822.

Neel, J. V., & Schull, W. J. Differential fertility and human evolution. In T. Dobzhansky, M. K. Hecht, & W. Steere (Eds.), *Evolutionary biology,* Vol. 6. New York: Appleton-Century-Crofts, 1972. Pp. 363–379.

Ralls, K. Auditory sensitivity in mice: *Peromyscus and Mus musculus. Animal Behavior,* 1967, **15,** 123–128.

Roberts, R. C. Some concepts and methods in quantitative genetics. In J. Hirsch (Ed.), *Behavior–genetic analysis,* New York: McGraw-Hill, 1967, Pp. 214–257. (a)

Roberts, R. C. Some evolutionary implications of behavior. *Canadian Journal of Genetic Cytology,* 1967, **9,** 419–435. (b)

Robson, G. C., & Richards, O. W. *The variations of animals in nature.* London: Longmans, Green, 1936.

Roe, A., & Simpson, G. G. (Eds.) *Behavior and evolution.* New Haven: Yale University Press, 1958.

Schull, W. J., & Neel, J. V. *The effects of inbreeding on Japanese children.* New York: Harper & Row, 1965.

Selander, R. K., & Kaufman, D. W. Genic variability and strategies of adaptation in animals. *Proceedings of the National Academy of Science, U.S.,* 1973, **70** 1875–1877.

Selander, R. K., & Yang, S. Y. Biochemical genetics and behavior in wild house mouse populations. In G. Lindzey & D. D. Thiessen (Eds.), *Contributions to behavior–genetic analysis.* New York: Appleton-Century-Crofts, 1970. Pp. 293–334.

Sidman, R. L., Green, M. C., & Appel. S. H. *Catalog of the neurological mutants of the mouse.* Cambridge, Mass.: Harvard University Press, 1965.

Thiessen, D. D. *Gene organization and behavior.* New York: Random House, 1972.

Thiessen, D. D., Owen, K., & Whitsett, M. Chromosome mapping of behavioral activities. In G. Lindzey & D. D. Thiessen (Eds.), *Contributions to behavior–genetic analysis.* New York: Appleton-Century-Crofts, 1970. Pp. 161–204.

Vandenberg, S. G. The nature and nurture of intelligence. In D. C. Glass (Ed.), *Genetics: Biology and behavior.* New York: Rockefeller University Press & Russel Sage Foundation, 1968. Pp. 3–58.

Vandenberg, S. G. What do we know today about the inheritance of intelligence and how do we know it? In R. Cancro (Ed.), *Intelligence: Genetic and environmental influences.* New York: Grune & Stratton, 1971. Pp. 182–218.

Wahlsten, D. Genetic experiments with animal learning: A critical review. *Behavioral Biology,* 1972, **7,** 143–182.

Wahlsten, D. Heritable aspects of anomalous myelinated fibre tracts in the forebrain of the laboratory mouse. *Brain Research,* 1974, **68,** 1–18.

Whitney, G. Vocalization of mice influenced by a single gene in a heterogeneous population. *Behavior Genetics,* 1973, **3,** 57–64.

Whitney, G., Alpern, M., Dizinno, G., & Horowitz, G. Female odors evoke ultrasounds from male mice. *Animal Learning & Behavior,* 1974, **2,** 13–18.

Whitney, G., Coble, J.R., Stockton, M. D. & Tilson, E. F. Untrasonic emissions: Do they facilitate courtship of mice? *Journal of Comparative & Physiological Psychology,* 1973, **84,** 445–452.

Whitney, G., McClearn, G. E., & DeFries, J. C. Heritability of alcohol preference in laboratory mice and rats. *Journal of Heredity,* 1970, **61,** 165–169.

Wimer, R. E., Wimer, C. C., & Roderick, T. H. Genetic variability in forebrain structures between inbred strains of mice. *Brain Research,* 1969, **16,** 257–264.

Wynne-Edwards, V. C. Population control and social selection in animals. In D. C. Glass (Ed.), *Genetics: Biology and behavior.* New York: Rockefeller University Press & Russell Sage Foundation, 1968. Pp. 143–163.

6

What Animals Should We Compare?

C. B. G. Campbell

Center for Neural Sciences
Indiana University[1]

The statement has sometimes been made that comparative neurology has contributed little or nothing of fundamental importance to an understanding of the nature of the nervous system and behavior. Such statements are often derived from human-oriented workers whose interests are primarily physiological or biochemical. Although such a statement is certainly debatable, it can be argued that comparative neurology has not yet realized its full potential as a scientific discipline. There are, I believe, a number of reasons why this may be so. I consider two of them to have been of some importance.

First, comparative neuroanatomists continued to pursue somewhat random, nonexperimental, descriptive studies at a time when more medically oriented workers were making important discoveries using newer experimental techniques. Second, the goals of comparative studies have often been unclear or have been concerned primarily with demonstrating increasingly complex levels of organization "as one ascends the phylogenetic scale" (to use an unfortunately common phrase). Comparative neurology, a discipline that should have been a bridge between the neurosciences and the broader field of zoology, has remained at a pre-World War II level of sophistication, whereas both zoology and the neurosciences have developed rapidly. In the past 10 or 15 years, comparative neurology has begun to use all of the modern physiological, anatomical, and biochemical technical methods currently available. This has resulted in some discoveries of great importance, many of which are discussed in this book. Unfortunately, not as much concern has been shown for the goals and methods

[1] Current Address: Departments of Radiological Sciences and Anatomy, University of California, Irvine, California.

of comparison. Such a concern should certainly be as central as that for using modern apparatus and technology.

Some morphological or functional condition of the nervous systems of different animals can be compared for any one of several reasons. The three I consider to be most significant are (1) to determine whether, and to what degree, the findings derived from the study of a single species can be generalized to the hierarchy of taxa to which that species belongs; (2) to study how the nervous system is involved in the adaptation of animals to varying modes of life; and (3) to attempt to infer the historical course of change in nervous system organization and function in the lineages that produced the animals living today. The animals chosen for comparison in any study are likely to be quite different depending on which reason for comparison is operating.

If the interest is in determining the generalizability of some feature, then animals of widely different taxa (i.e., of different lineages) are chosen and they should be adapted to widely varying modes of life. An example of such a study is that by Moore (1973), in which he infers that the retina of the eye sends optic nerve fibers directly to the suprachiasmatic nuclei of the ventral hypothalamus in mammals as a group. Retinohypothalamic connections in mammals have been denied by some previous workers (e.g., Kiernan, 1967) and affirmed by others (e.g., Campbell, 1969).

If the effects on the nervous system of adaptation to an arboreal mode of life are being studied, then animals that live in the trees but are not all closely related are chosen. Such animals may be compared as arboreal monkeys, arboreal squirrels, tree shrews, arboreal rodents other than sciuromorphs, arboreal carnivores (e.g., the kinkajou), and the marsupial phalanger. These animals, of course, would have to be compared with nonarboreal mammals to be sure that the feature was lacking in such animals. Studies of this sort have been very rare and those that exist usually compare only two or three animals. This is unfortunate, for of the three types of comparative studies listed above this type may be expected to have the most real significance for students of function.

Historical inference requires the most careful consideration of animal choices. It is perhaps in this realm that the greatest number of errors have been made. It was commonplace in the past to compare some feature in the morphological sequence of rat, cat, monkey, and man and to infer that this represented a historical sequence. Fortunately, this practice is less common today. A distinct improvement has been the recent emphasis on making comparisons within lineages (Hodos & Campbell, 1969). Some groups of workers have been making such comparisons. This new emphasis has produced some very interesting and important findings but, because the method followed has not always been entirely consistent with prevailing zoological practice, it has not been as reliable as it may otherwise be.

Warren (1973) indicates that the suggestion that many of the difficulties inherent in comparing species from unrelated taxa can be avoided by making comparisons within lineages does not seem to hold true in a study by Glickman and Sroges (1966). These workers studied strength of curiosity in various species of primates and carnivores and found that this feature was related to feeding habits and species-typical patterns of evading predators rather than to their "taxonomic standing" within their order (presumably Warren refers to whether or not they are "primitive" or "advanced"). As Hodos and Campbell (1969) stated, and as I have reiterated above, the comparisons within lineages were for the purpose of inferring historical change within a lineage. The Glickman and Sroges (1966) study is an example of the analysis of adaptation, also discussed by Hodos and Campbell (1969), in which confining comparisons to a single lineage is not indicated. It should be understood that I use the term "historical" to mean what did occur over a period of time.

The chief difficulty with existing studies making comparisons within lineages is that the comparisons are not extensive enough to determine which features are truly ancestral (primitive) or derived (specialized) within the lineage under consideration. To illustrate the problem I shall discuss a recently published hypothesis on the course of evolution of the central visual pathways in the primate lineage (Diamond & Hall, 1969). This hypothesis describes four main historical steps suggested by studies on three living groups of animals and one hypothetical construct (Figure 1). Figure 1a represents a hypothetical reptilelike mammal in which optic axons project from the eye to the optic tectum only. The optic tectum then projects to an undifferentiated thalamic visual nucleus, which transmits visual information to a general sensory cortex. The second step (b) is represented by the living tree shrew, *Tupaia glis,* in which the eye projects to both the thalamus and the optic tectum (Figure 1b). The thalamic optic nucleus is differentiated into two nuclei: one that receives input directly from the eye (GL), and one that receives its input indirectly via the optic tectum (LP). The projections of these two nuclei to the visual core cortex and visual belt cortex overlap. The third stage (c), represented by living prosimians (Figure 1c), and the fourth (d), represented by living anthropoid primates (Figure 1d), demonstrate progressive differentiation and separation of the two pathways to the visual cortex.

Studies on the visual pathways of living reptiles (Butler & Northcutt, 1971; Hall & Ebner, 1970) and living reptilelike mammals (Campbell & Hayhow, 1971, 1972) suggest that reptilelike mammals have also had direct retinal projections to thalamic nuclei. The tree shrew, used to represent Stage b, is not now believed to be a member of the primate lineage (Campbell, 1965, 1966a, b; Martin, 1968; McKenna, 1966; Van Valen, 1965; see also Radinsky, 1976, Volume I, Chapter 12). Consequently the organization of the tree shrew visual system probably does not represent a stage through which the primate lineage has passed. Only c and d remain clearly as members of the primate lineage. New information can of course invalidate any model or hypothesis, so Diamond and Hall's attempt to describe the evolution of the primate visual system cannot be faulted on this

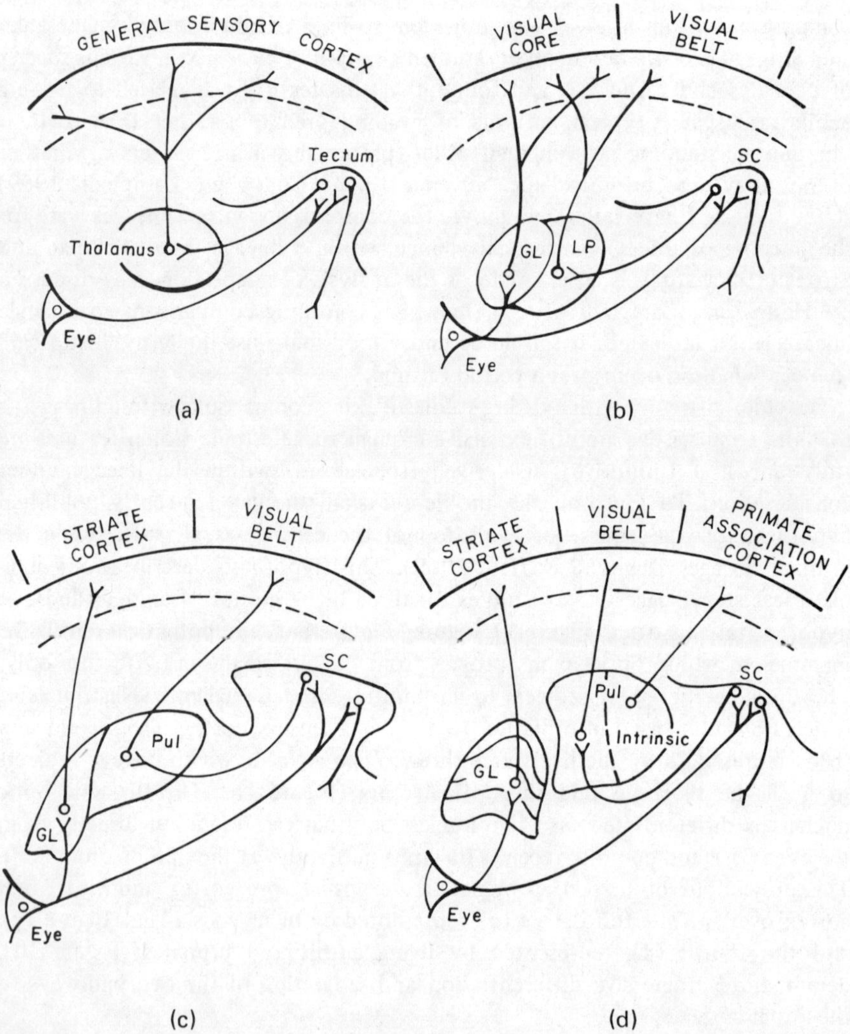

FIG. 1 Evolution of the central visual pathways in primates, according to Diamond and Hall (1969): (a) reptile-like mammal; (b) tree shrew; (c) prosimian primate; (d) anthropoid primate.

ground. However, their interesting hypothesis might have remained useful even longer if the stages described had resulted from more complete analysis of ancestral (primitive) and derived (specialized) character states in the visual system of primates.

To make a valid inference of the history of the primate central visual pathways, living members of the primate lineage would have to be adequately sampled. If several different patterns of organization emerge from such a study, then the next step is to attempt to determine which of the patterns is ancestral and which are derived within this lineage. It must be remembered that a morphologically simple pattern may not be the ancestral one but may result from secondary simplification. Placing the patterns of organization within the framework of the currently accepted phylogeny of the primate order can very likely suggest which pattern is ancestral and which patterns are derived. This can be confirmed or possibly rejected on the basis of comparisons with other lineages derived from the same common ancestry as that of primates.

The logic is illustrated in Figure 2. Suppose interest is primarily in the lineage derived from the now extinct ancestor B and represented by the extant species F, G, H, I, and J. These animals are examined and two characters are found, represented by the solid squares and the open circles. How can which character is ancestral and which derived be determined in this lineage? If the pattern of temporal relationships shown in the dendrogram is known, then that is of some help. However, examining two other lineages that share, with B, a common ancestry with the extinct A can provide an added clue. In addition to the characters represented by the solid circles and triangles and the shaded triangles, each of these lineages have representatives with the same character (solid square) found in lineage B. If parallelism and convergence can be ruled out (Campbell & Hodos, 1970) then the most likely inference is that the character represented by the solid square is present in both B and A. In this way the conclusion can be reached that "solid square" is the ancestral character and that the other characters are derived. Knowledge of the phylogeny of the groups is essential, and comparisons within lineages are still primary. Comparisons between different lineages derived from a common ancestry, and here intended for a specific purpose, are not the same as the old rat–cat–monkey–man comparisons.

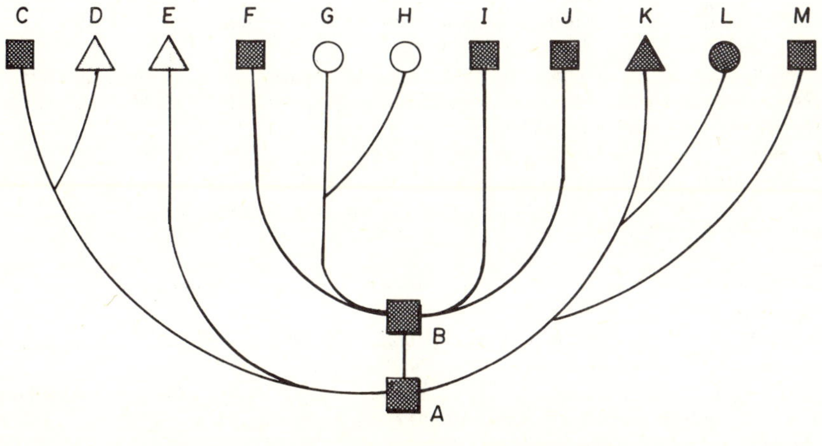

FIG. 2 Dendrogram of a hypothetical phylogeny. See text.

This is, of course, a highly simplified example. The construction of valid historical sequences is not easy and requires considerable information. Some of this information is very difficult to get, especially if it requires extensive experimental study in each species examined. It is certainly tempting to construct hypothetical phylogenies when a reasonable morphological or functional sequence is present in three or four species of the same lineage. It is possible that, by chance, the correct historical sequence can be inferred; however, the possibility of error is strong.

Not every student of comparative neurology will be concerned with the active pursuit of actual historical sequences. There are many whose interests are primarily physiological or biochemical, and their choice of animals for study may depend on considerations other than those reviewed here. No matter what the aim of a proposed study may be, however, the choice of animals to be compared must be made with care and within a conceptual framework compatible with zoological principles. Only in this way can comparative neurology make valid and significant contributions to the study of animal life.

REFERENCES

Butler, A. B., & Northcutt, R. G. Retinal projections in *Iguana iguana* and *Anolis carolinensis.* Brain Research, 1971, **26,** 1–13.

Campbell, C. B. G. "Pyramidal tracts in primate taxonomy." Ph.D. Thesis, University of Illinois, 1965.

Campbell, C. B. G. Taxonomic status of tree shrews. *Science,* 1966a, **153,** 436.

Campbell, C. B. G. The relationships of the tree shrews: the evidence of the nervous system. *Evolution,* 1966b, **20,** 276–281.

Campbell, C. B. G. The visual system of primates and insectivores. *Annals of the New York Academy of Sciences,* 1969, **167,** 388–403.

Campbell, C. B. G., & Hayhow, W. R. Primary optic pathways in the echidna, *Tachyglossus aculeatus:* an experimental degeneration study. *Journal of Comparative Neurology,* 1971, **143,** 119–136.

Campbell, C. B. G., & Hayhow, W. R. Primary optic pathways in the duckbill platypus, *Ornithorhynchus anatinus:* an experimental degeneration study. *Journal of Comparative Neurology,* 1972, **145,** 195–208.

Campbell, C. B. G. & Hodos, W. The concept of homology and the evolution of the nervous system. *Brain, Behavior, & Evolution,* 1970, **3,** 353–367.

Diamond, I. T., & Hall, W. C. Evolution of neocortex. *Science,* 1969, **164,** 251–268.

Glickman, S. E., & Sroges, R. W. Curiosity in zoo animals. *Behavior,* 1966, **26,** 151–188.

Hall, W. C., & Ebner, F. F. Parallels in the visual afferent projections of the thalamus in the hedgehog (*Paraechinus hypomelas*) and the turtle (*Pseudemys scripta*). *Brain, Behavior, & Evolution,* 1970, **3,** 135–154.

Hodos, W., & Campbell, C. B. G. *Scala naturae:* Why there is no theory in comparative psychology. *Psychological Review,* 1969, **76,** 337–350.

Kiernan, J. A. On the probable absence of retino-hypothalamic connexions in five mammals and an amphibian. *Journal of Comparative Neurology,* 1967, **131,** 405–408.

McKenna, M. C. Paleontology and the origin of the primates. *Folia Primatologica,* 1966, **4,** 1–25.

Martin, R. D. Towards a new definition of primates. *Man,* 1968, **3,** 377–401.

Moore, R. Y. Retinohypothalamic projection in mammals: a comparative study. *Brain Research,* 1973, **49,** 403–409.

Radinsky, L. B. Later mammal radiations. In R. B. Masterton (Ed.), *Evolution of the nervous system and behavior*. Hillsdale, New Jersey: Lawrence Erlbaum Assoc., 1976.

Van Valen, L. Treeshrews, primates and fossils. *Evolution,* 1965, **19,** 137–151.

Warren, J. M. Learning in vertebrates. In D. A. Dewsbury & D. A. Rethlingshafer (Eds.), *Comparative psychology: A modern survey*. New York: McGraw Hill, 1973. Pp. 471–509.

7
Grades and Clades Revisited

Stephen Jay Gould

Harvard University

INTRODUCTION

I have a simple faith that the best people in any field are never moronic and are hardly ever truly misguided. Therefore, in the face of acrimonious debate, I generally assume that semantic difficulties inspire 90% of any argument and that, when these are sorted out, both sides are doing something right.

The greatest difference in method constantly before us in this symposium pits paleontologists against students of comparative neurology and behavior. The paleontologists are engaged in constructing ever more finely tuned phylogenies; the simple ladders of an earlier century have yielded to bushes of the most luxurious and complex branching. In the face of this, it seems as though comparative neurologists remain rooted to Lamarck's *scala naturae*—for they persist in studying a fish, a reptile, an insectivore, a tree shrew, a monkey, and a man and in drawing from such comparisons a set of conclusions about vertebrate evolution. The neuroanatomists speak of their sequence as though it represents phylogeny, which it manifestly does not. Paleontologists then seem to assert that because the series does not reflect true descent, it cannot designate anything of value. I propose (1) that this debate involves words and not things; (2) that it will be laid to rest when neuroanatomists cease implying that their procedure is related to true phylogeny; (3) that this procedure, nonetheless, is both valid and valuable; and (4) that it reflects an evolutionary concept of undoubted respectability, despite its inevitably subjective basis. In short, I assert that the linear sequences of neuroanatomy are successive grades rather than phyletic clades and that analysis by grades is both a neglected and an important mode of evolutionary comparison.

THE ELUSIVE NOTION OF EVOLUTIONARY "PROGRESS"

"Never say 'higher' or 'lower,' " Darwin exhorted himself; for the temptation remained even though evolution by natural selection guaranteed only adaptive change to environmental conditions, not an ineluctable march toward any vision of organic perfection. Nonetheless, Darwin (1859) recognized that natural selection would yield a sort of "progress"; most adaptations are specific responses to specific conditions (eyeless moles and long-necked giraffes) but the more general advantage that improved design provides will have its cumulative effect through time:

> The inhabitants of each successive period in the world's history have beaten their predecessors in the race for life, and are, in so far, higher in the scale of nature; and this may account for that vague yet ill-defined sentiment, felt by many paleontologists, that organization on the whole has progressed [p. 345].

Because we are not willing to abandon the notion of "progress," we shall have to define it, hopefully in a minimally ambiguous and anthropocentric way. Because it is always easier to specify what a thing is not, I shall state the cardinal principle of any adequate definition of evolutionary progress: it must not be defined by phylogeny—an animal is not "higher" simply because it has evolved later. Phylogeny is a (potential) fact that must be kept separate from interpretations of its meaning. If progress is defined by position in a branching sequence, it becomes a vacuous notion with no content independent of the phyletic tree itself. (High levels of empirical concordance between successive grades and temporal positions in a true phylogeny should not be surprising, however; animals of better design are, after all, likely to prevail in competition with their forebears.)

I think that the most useful concept of evolutionary progress must be linked to design criteria and optimization in the engineering sense. This imposes great restrictions on my hopes: (1) I shall be hard put to apply it to those important phenomena—intelligence, for example—that are so hard to quantify or specify in mechanical terms; (2) I must know how a structure is used throughout its history, for a marked shift in function resets the optimal form (in what sense can a tetrapod limb be "better" than a crossopterygian fin?); (3) because major changes in environment mark the history of most major groups, analyses must proceed in sections (swimming vertebrates separately from terrestrial tetrapods, for example).

Moreover, since evolution follows a patchwork, or mosaic, course, stages of progress (or grades) differ according to the criterion employed. Fishes, analyzed as swimming machines, form different grades than do the same fishes judged as feeders. Phylogeny has the undoubted advantage of its potential objectivity—there is a true branching sequence, even if it may never be known. Grades, stages of progress, or levels of organization vary according to the criteria employed, the organs analyzed, and the limits of current knowledge about optimal design and the nature of materials. Nonetheless, such an attractive notion cannot be avoided just because it cannot be endowed with the objectivity of the evolutionary tree. A teleost fish is a better swimming machine than a paleoniscoid; a lion leaping on a zebra represents something beyond *Dimetrodon* waddling after *Edaphosaurus.*

Grades, then, are successive levels of organization defined as stages in the improvement of an organic design for some specified function. There is nothing mystical or orthogenetic in a temporal sequence of grades in the history of a major group; to engender such a sequence, recourse need be had to little more than the evolutionary platitudes that (1) better designs tend to be favored by natural selection and (2) animals entering new adaptive zones are generally suboptimal in design because of constraints of heritage (early tetrapod limbs project laterally because crossopterygian fins did so for good aquatic reasons). Clades, in contrast, are the true monophyletic branches of an evolutionary tree (clades may move from one grade to the next or, more commonly, develop entirely within a given grade).

Many great evolutionists have had a "gut" feeling–and I trust gut feelings with the same simple faith that motivated my opening sentence–that diversification within a grade is an event of a different order from transitions between grades. This feeling has generated a host of terms–aromorphosis versus idioadaptation (Severtsov, 1931), and anagenesis versus cladogenesis (Rensch, 1959; Huxley, 1958, for example). Even Lamarck (1809, pp. 130–137) made a similar distinction between the "cause which tends incessantly to complicate organization" and "the influence of circumstances." The subsequent analysis follows Huxley's examples but uses my restricted definitions.

THE ANALYSIS OF GRADES AND CLADES

The study of allometry, or relative growth, has been Julian Huxley's finest contribution to evolutionary theory (Huxley, 1932; Gould, 1966). In developing a mathematics for allometry, his cardinal insight was the recognition that time could be bypassed by plotting one dimension against another. (The standard curves of growth vs. time are complex functions with parameters that have resisted biological interpretation; the simple power function, or "allometric equation" of one part versus another, can be calculated and interpreted by a tyro; Figure 1). With this success firmly in mind, Huxley decided to pursue an analogous course in the depiction of evolution. The standard evolutionary tree is not a full, objective, and inevitable rendering for the history of life; it is a highly specific plotting device that puts time on the ordinate and cladistic divergence on the abscissa and that leaves out evolutionary progress (or anagenesis) entirely. Huxley decided, instead, to bypass time and to plot sequences of grades

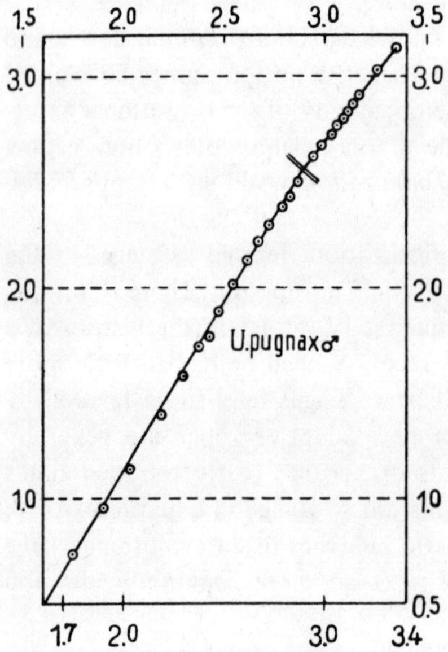

FIG. 1 Claw weight vs. body weight in the ontogeny of male fiddler crabs: Huxley's classic example of allometry as a bypass of time (from Huxley, 1932). With strong positive allometry, old and large crabs have relatively larger claws. This simple relation is not at all evident when both variates are plotted directly against time.

(anagenesis) on the ordinate vs. diversification (or cladogenesis) on the abscissa. Figures 2 and 3 reproduce two of his typical plots. They display both the defining criteria and the empirical characteristics of evolutionary grades.

Subjective Definition

Each series of grades reflects progress according to a stated criterion and for a specific organ or part. Figure 3, for example, represents a reasonable sequence of grades for progress in the evolution of the primate brain defined by increasing relative size at corresponding body weights. The same sequence becomes totally inadequate for an analysis of locomotion: quadrupedal ground walking, branch walking, brachiation, and bipedal ground walking are simply different, equally efficient ways of moving in different environments.

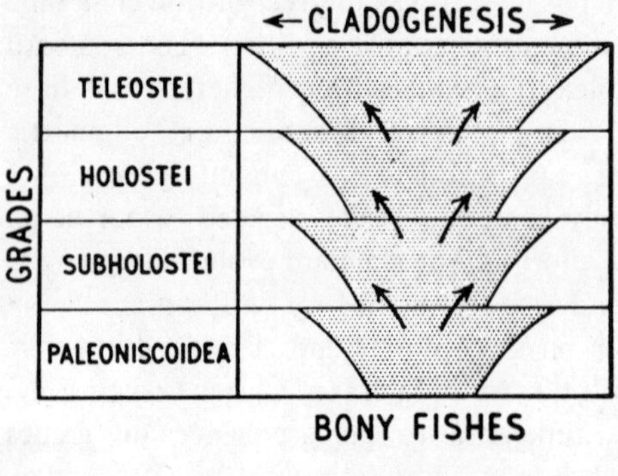

FIG. 2 A rough system of grades for the evolution of actinopterygian fishes (from Huxley, 1958), based on criteria of swimming and locomotion. Most modern students would now advocate a monophyletic origin for teleosts.

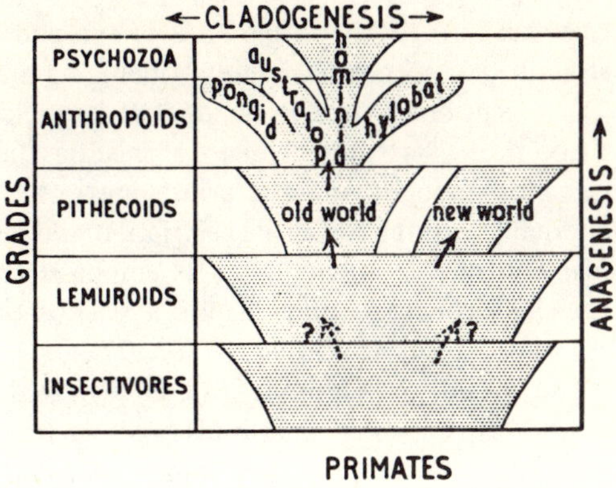

FIG. 3 A rough system of grades for the evolution of primates based on increasing brain size for given body weights. (From Huxley, 1958.)

Constancy of Environment and the Engineering Criterion of Progress

If Figure 2 is construed as a sequence of grades for hydrodynamic efficiency in swimming, then the general habitat, pelagic swimming, can be taken as reasonably constant and environmental shifts affecting the evolution of other organs can be ignored. It must be argued that paleoniscoids were suboptimal swimmers. A Devonian engineer, asked to design a better swimming machine with the same basic materials, could have specified the blueprints for a series of improved designs. Evolution, when operating in broad and relatively stable environments, should display some predictable components. D'Arcy Thompson (1942) often spoke of "the comparatively few avenues of physico-mathematical possibility." Starting with the paleoniscoid design, the paths to improvement were few in theory and well trodden in practice. The characters defining the grades in Figure 2 include evolution of the air bladder, development of a homocercal tail, and forward migration of the pelvic fins. All provide increased efficiency in swimming on purely mechanical criteria (Harris, 1936; Alexander, 1968).

Polyphyly, Parallel Evolution, and the Analog

If efficient solutions to common problems of optimization are as limited as D'Arcy Thompson has claimed, then grades are levels of structural organization that may be reached independently by different lineages. Many higher taxa are not the monophyletic clades that most theories of classification require (or at least desire) but are grades of improvement attained in the same way by many lineages. The common features that unite the clades of such taxa are not homologous but only analogous. When solutions to common problems are limited and success confers great advantages on lineages in competition, parallel evolution is rampant. A "phyletic bias"—and its metaphor, the evolutionary tree—has prevented proper assessment and understanding of the severe constraints that mechanical limits (including size itself) place on adaptive design. The evolutionary tree is no neutral hatrack for the facts of evolution; it is a theoretical construct tied to a specific theory of organic resemblances and directing thought along a priori lines. The primary value of Huxley's (1958) alternate picture, for all its restriction and subjectivity, is its frank display of a bias contrary to orthodoxy. Its emphasis on analogous resemblance and functional morphology may inspire a science of form as powerful as the current science of diversity (Gould, 1970, 1971).

The order of polyphyly can be removed by definition. If monotremes are judged as surviving therapsids and such a category as MacIntyre's "quasi-mammals" is adopted for some inconvenient progenitors, then marsupials, placentals, and their immediate ancestors form a monophyletic taxon. This is a word play that preserves current preferences but confuses a great issue in the evolution of form. We can designate and divide it in any way we wish, but the fact remains that many therapsid lineages independently evolved an impressive suite of features in concert, including such intricate transformations as the conversion of jaw bones to ear bones. Olson, in this volume, has listed an impressive series of rapid innovations occurring in several lineages and involving such structures as the floccular fossa, the forebrain, the semicircular canals, and the pineal gland.

Patterns in Diversity

Improvement in design should have general consequences beyond the approach to optimal form itself. Increasing cladistic diversity in successive grades may be predicted from the general supposition that broad improvements in design increase the range of potential variations on it (homoiothermy as a prerequisite for true flight among vertebrates, for example). (The failure of this prediction in Figure 3 robs the prediction of any universality; in this case, locomotory and other specializations accompanied the general increase in cranial capacity and limited the potential scope of diversification—the prediction works better for grades constructed at higher taxonomic levels.) Mammals and teleosts, as the highest grades of aquatic and terrestrial vertebrates, respectively, are both more abundant in species and more widely diversified in form and habitat than are other vertebrate grades. Jerison, in this volume and elsewhere (1970, 1973), has shown that successive grades in the evolution of the mammalian brain are marked not only by larger mean brain weights at standard body weights but also by increasing ranges of diversity in brain size and function. In the most comprehensive example of all, Whittaker's "five kingdom system" (1969) arranges the living world into three great grades: procaryotic protists (the Monera), eucaryotic protists (the Protista), and eucaryotic multicellular organisms (the Plantae, Fungi, and Animalia). Diversity in the highest grade is so pronounced that it is recognized by establishing three vertical divisions within a single level. To emphasize a previous point, not only was multicellularity attained independently in three distinct ways, it was almost surely attained many times within each of the three multicellular grades.

WHY ANALYZE BY GRADES?

Faced with the dilemma of their subjective status, why should evolutionists analyze the history of a group through its grades if they can also do so according to its true phylogeny? This question inspires two separate responses.

First, "if they can also do so" represents a hope more vain than most nonpaleontologists imagine. Every field has its trade secret: paleontology's is that the true phylogenies of exceedingly few groups are known. The imperfection of the fossil record and the nature of evolution itself combine to guarantee that this lack of information is no result of insufficient effort. Darwin compares the fossil record to a book retaining only a letter or two per page; in addition, evolutionary change almost always proceeds by rapid events of speciation in small, isolated populations, not by slow steady transformations of entire species (Eldredge & Gould, 1972). These two statements provide little comfort to those who hope to read evolution from the rocks. Fossils generally represent the termini of a few successful branches, not the progenitors of anything. Of course, even if phylogenies could be constructed to rival Haeckel's fondest dreams, this would provide no help for comparative neurologists, who must study living organisms.

Second, I do not think that analysis by grades is likely to be abandoned even in a pipedream world where phylogenies are laid out upon laboratory tables. To measure the hydrodynamic efficiency of an ichthyosaur, should it be compared with a fish or with its true terrestrial ancestor? To measure the capacity of a pterodactyl for flight, should it be contrasted with a bird or with its bipedal reptilian progenitor? The analog, as I have emphasized elsewhere in this volume (pp. 175–179), is a proper object for evolutionary comparison when we study what an organ does rather than where it came from. Lemur–monkey–ape–man is a caricature of primate phylogeny; once we are sure that the species we study are representative of their grade, this same sequence may well unravel the mysteries of neocortical function in successive levels of organization of the primate brain. Long live the analog and its science of functional morphology; and, with apologies to Menander, call a grade a grade.

REFERENCES

Alexander, R. M. *Animal mechanics.* Seattle: University of Washington Press, 1968.
Darwin, C. *On the origin of species.* London: John Murray, 1859.

Eldredge, N., & Gould, S. J. Punctuated equilibria: an alternative to phyletic gradualism. In T. J. M. Schopf (Ed.), *Models in paleobiology*. San Francisco: Freeman, Cooper, 1972. Pp. 82–115.

Gould, S. J. Allometry and size in ontogeny and phylogeny. *Biological Reviews,* 1966, **41,** 587–640.

Gould, S. J. Evolutionary paleontology and the science of form. *Earth Science Review,* 1970, **6,** 77–119.

Gould, S. J. D'Arcy Thompson and the science of form. *New Literary History,* 1971, **II**(2), 229–258.

Harris, J. E. The role of the fins in the equilibrium of the swimming fish, *Journal of Experimental Biology,* 1936, **13,** 476–493.

Huxley, J. S. *Problems of relative growth.* London: MacVeagh, 1932.

Huxley, J. S. *Evolutionary processes and taxonomy with special reference to grades.* Uppsala Univ. Arsk., 1958. Pp. 21–39.

Jerison, H. J. Brain evolution: new light on old principles, *Science,* 1970, **170,** 1224–1225.

Jerison, H. J. *Evolution of the brain and intelligence*. New York: Academic Press, 1973.

Lamarck, J. P. B. *Philosophie zoologique*. Paris: Dentu, 1809.

Rensch, B. *Evolution above the species level*. New York: Columbia University Press, 1959.

Severtsov, A. N. *Morphologische gesetzmässigkeiten der evolution*, Jena: 1931.

Thompson, D. W. *On growth and form*. Cambridge, England: Cambridge University Press, 1942.

Whittaker, R. H. New concepts of kingdoms of organism, *Science*, 1969, **163**, 150–160.

8

Interrelationships of the Major Fish Groups with Comments on Specimen Availability

Bobb Schaeffer

American Museum of Natural History

The fossil record for the fishes extends back to the Ordovician Period, nearly 500 million years ago (see Moy-Thomas & Miles, 1971; Norman & Greenwood, 1963). By that time, morphogenetic systems had been established for the deposition of cartilage, bone, dentine, and an enamel-like substance now usually called "enameloid." The first vertebrates were presumably jawless forms; by the Silurian Period they were represented by several different kinds of ostracoderms (Osteostraci, Anaspida, and Heterostraci; see Figure 1). The living lampreys (petromyzontoidei) and hagfishes (Myxinoidei) show certain resemblances to some of the ostracoderms, but there is no firm evidence regarding their relationships. The petromyzontoids are modified for parasitic feeding, whereas the myxinoids are adapted to carrion and bottom feeding. Their specializations should be kept in mind in any consideration of their brain organization.

Hagfishes are cold-water marine animals and they are usually trapped offshore. They are available along the Pacific coast of North America and on the Atlantic south of the Canadian Maritime Provinces. Lampreys occur in cold, fresh water. State or university biological surveys should be consulted in regard to obtaining specimens.

Two major groups of jawed bony fishes, the acanthodians and the placoderms, have been extinct for millions of years and are, therefore, of little direct interest to neurologists. Acanthodian endocasts are unavailable, but it has been possible to prepare placoderm endocasts from three-dimensional specimens. (see Schaeffer, 1969).

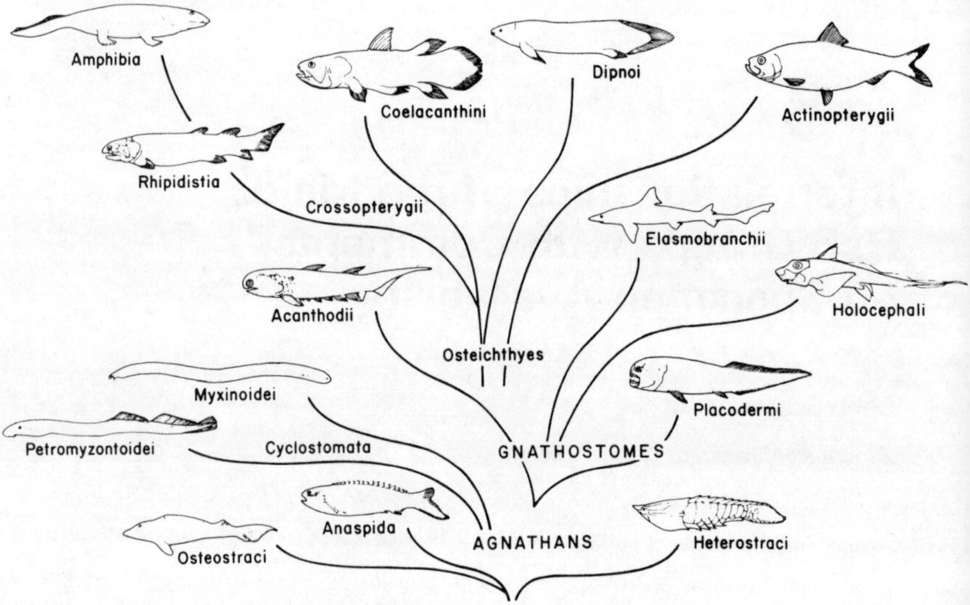

FIG. 1 Phylogeny of the major groups of fishes.

The large group of fishes generally called the Chondrichthyes includes the elasmobranchs (sharks, skates, and rays) and the holocephalans (chimaeras, or ratfishes). The two groups share certain characters, such as claspers in the male, that indicate common ancestry. They have, however, existed as separate groups since the Devonian.

The most primitive living sharks are the hexanchids (cow sharks, or six- or seven-gilled sharks), the chlamydoselachids (frilled sharks) and the heterodontids (bullhead sharks). The hexanchids are worldwide in distribution and occur in waters that range from shallow to quite deep. The frilled shark (*Chlamydoselache*) is a rare deep-water form, occasionally captured in the Pacific and more rarely in the Atlantic. The bullhead sharks, including the Port Jackson shark and the horn shark, are essentially Pacific forms and are unknown in the Atlantic or Mediterranean. Two species of cow sharks and one of the horn shark occur along the Pacific coast of the United States, but not in predictable numbers. Only the latter has been kept successfully in captivity. In fact, very few elasmobranchs are hardy in captivity. Marine stations and aquaria should be approached for live material.

The holocephalans are represented today by a few rather rare genera of chimaeroids. Specimens of *Hydrolagus*, which occurs in both the Atlantic and the Pacific, are frequently obtained in Puget Sound. The chimaeras are more delicate than the sharks. Friday Harbor Laboratories at the University of Washington has been able to keep them alive for limited experimental studies.

The higher bony fishes, or osteichthyans, are usually divided into two groups: the actinopterygians, or ray-finned fishes, and the sarcopterygians, or lobe-finned fishes.

The sarcopterygians include the crossopterygians (rhipidistians and coelacanths) and the dipnoans (lungfishes). The coelacanth *Latimeria* is the only living representative of the crossopterygians. The coelacanths have persisted with little change for about 400 million years. They are related to, but quite distinct from, the extinct rhipidistians that, in turn, are related to the amphibians. Various characters in *Latimeria*, particularly in the soft anatomy, are difficult to evaluate because of its isolated systematic position. An average of three coelacanths has been caught each year since 1938 in the Comoro Islands. Enquiry concerning their availibility should be addressed to the Laboratoire d'Anatomie comparée du Muséum national D'Histoire naturelle, 55, rue de Buffon, 75005 Paris.

The conservative dipnoans presumably shared with the crossopterygians an Early Devonian or Silurian ancestor. Of the three living genera (*Lepidosiren, Protopterus,* and *Neoceratodus*), only *Neoceratodus* from Australia is closely related to the Mesozoic *Ceratodus*. The ancestry of the other two genera remains problematical but, in any case, it is apparently far removed from that of *Neoceratodus*. Because the dipnoans have had a long independent history, it is of particular interest that they show numerous amphibian-like characters in their soft anatomy and physiology. Some workers feel that the resemblances are the result of convergence, but it is probable that many of them are primitive sarcopterygian characters. *Latimeria*, however, shows few resemblances to the tetrapods.

The Australian lungfish, *Neoceratodus*, can be obtained only with the special permission of the Australian government. The African *Protopterus* and the South American *Lepidosiren* are sometimes available through the larger tropical fish dealers but there is usually a delay in filling orders.

By far the most successful osteichthyans are the ray-finned fishes, or actinopterygians. Their record also begins in or before the Devonian, and today there are more species of actinopterygians than of all other vertebrates together. They are frequently divided into the chondrosteans (including the extinct palaeonisciforms), the holosteans, and the teleosts, but the first two groups, at least, are probably paraphyletic (see Greenwood *et al*., 1973).

Living chondrosteans are the sturgeons, paddlefishes, *Polypterus,* and *Calamoichthys*. All of these forms extend back to unknown palaeonisciform acestors. The gar (*Lepisosteus*) and the bowfin (*Amia*) are living holosteans. Their lineages have probably been independent for over 200 million years.

Polypterus and *Calamoichthys* may be difficult to obtain from tropical fish dealers because recent collecting operations in Africa have been restricted. Sturgeons, paddlefishes, gars, and bowfins require experienced collectors to obtain them. Such people generally belong to state fish and game departments, public aquaria, or fishery institutes of larger universities. The paddlefish is extremely difficult to maintain in captivity. Like the sturgeon, it requires large tanks and copious amounts of water. The bowfins and gars are relatively hardy but like all fishes they have limited homeostatic mechanisms, which make them delicate in captivity. Consultation with aquarium experts is recommended before experimental work with these fishes is attempted.

The most advanced of the actinopterygians are the exuberantly diversified teleosts, which include most of the bony fishes living today. They inhabit nearly every aquatic environment available to organisms. The systematics of the teleosts are unsettled, and major revisions in the classification may be expected. Anyone concerned with teleost neurology should refer regularly to the bibliographies in the *Zoological Record* for systematic revisions (see also Greenwood *et al.*, 1966).

The hardiest fishes under the conditions that can be maintained in most laboratories are those from tropical fresh waters; they include the lungfishes and polypterids. The outstanding exception is the goldfish, the most adaptable of all temperate species. Many different species of tropical freshwater teleosts are regularly available at pet stores (see Sterba, 1962).

Accurate identification of many fishes may be difficult. When in doubt, a professional ichthyologist should be consulted (see Membership Directory of the American Fisheries Society, 1972). If rare species are desired, or if examples of the species in a particular genus of family are required, their availibility can be determined most easily by consulting ichthyologists working in natural history museums or at major aquaria (see Fraga *et al,* 1972; Vernberg, 1963), as well as those on the staffs of local, state, and federal biological surveys.

ACKNOWLEDGMENTS

I wish to acknowledge the kindness of Dr. James Atz for supplying information on availibility of specimens.

REFERENCES

American Fisheries Society. Membership directory. American Fisheries Society, Miscellaneous Publications, 72–8, 1972.

Fraga, M. G., deLuca, C., & Tinker, S. *Directory of the public aquaria of the world, 1971–1972.* (4th ed.) Honolulu: Waikiki Aquarium, 1972.

Greenwood, P. H., Miles, R. S., & Patterson, C. (Eds.) *Interrelationships of fishes*. London: Academic Press, 1973. Supplement no. 1 to the *Zoological Journal of the Linnean Society*, **53**.

Greenwood, P. H., Rosen, D. E. Wietzman, S. H., & Myers, G. S. Phyletic studies of teleostean fishes, with a provisional classification of living forms. *Bulletin of the American Museum of Natural History,* 1966, **131**, 339–456.

Moy-Thomas, J. A., & Miles, R. S. *Palaeozoic fishes*. (2nd ed.) Philadelphia: W. B. Saunders, 1971.

Norman, J. R., & Greenwood, P. H. *A history of fishes.* (2nd ed.) New York: Hill & Wang, 1963.

Schaeffer, B. Adaptive radiation of the fishes and the fish-amphibian transition. *Annals of the New York Academy of Sciences,* 1969, **167**, 5–17.

Sterba, G. Freshwater fishes of the world. (Translated and revised by Denys W. Tacker.) London: Vista Books, 1962.

Vernberg, F. J. (arranger). Field stations of the United States. *American Zoologist*, 1963, **3**, 245–450.

9

The Nomenclature of Correspondence: A New Look at "Homology" and "Analogy"

Michael T. Ghiselin

The University of California, Berkeley

This chapter deals primarily with what is meant by the terms "homology" and "analogy," but it does not review the large and scattered literature on that topic. Rather, it seeks to put the critical issues into a new perspective, one which hopefully will help us to decide which of a number of possible usages is to be preferred.

In passing judgment on such matters, we should come to grips with some matters of formal logic and analytical philosophy that underlie the controversies, for the "semantic arguments" turn out to involve more than one might have expected. There are times when one should argue about the meaning of words, and there are times when one should not. So long as those who use a word understand one another, they can communicate effectively. They need not reflect on the precise meaning of terms, nor need they be able to supply anything like a dictionary definition, much less a philosophical analysis. We can say much the same for the language of our daily life, as well as for that of scientific discourse. Children use a wide variety of words, and quite effectively, long before they can define them. By the same token, we should not worry if a scientist has a hard time explaining the more recondite terminology. One rightly looks askance at those who criticize our use of words like "adaptation" because they seem, upon superficial analysis, paradoxical (Ghiselin, 1966a). On the other hand, language can be ambiguous and confusing, and communication becomes difficult when someone uses a word in one sense while his audience thinks he intends another. When we do encounter such difficulties, it helps to examine our language. We need to analyse our concepts and define our terms.

Yet how should we proceed with this business of definition? And just what is it that we seek to define–a word, a thing, or perhaps a concept? Should a definition be short, terse, and to the point, or can it run book length? Philosophers have written extensively upon such topics, and the "theory of definitions" forms a minor, albeit important, branch of logic (Robinson, 1954). Consensus, even as to how one defines "definition," has been long in coming, and even now there is room for argument. I myself prefer the view of Wittgenstein (1961, p. 61) that "A definition is a rule dealing with signs." The basic point here is that when we define we are making known the conventions that are established for the use of symbols, including words. According to this point of view, one that can indeed be contested, there is no ultimate court of appeal as to how words and other symbols must be used ("=" could mean "+" for example). We can coin terms at will and use them in new ways if we feel like it. The meaning of words is conventional. In going along with this thesis, we can leave out such interesting yet peripheral topics as to whether we can define something other than a symbol. At least we define our terms, rather than our things, and we can do so without having to worry about what it is that we do with our concepts. It makes a real difference, too. In some philosophical systems, such as those of Plato and Aristotle, it was assumed that we define the *essences* of things–the Ideas in the Divine Mind, or their equivalent in the things themselves, which supposedly accounted for the things being what they are. Such a view treats definitions as more than conventions. It has created problems, not only for biologists of the distant past, but for us moderns as well. To give an example, Simpson (1966) criticized von Frisch for referring to the "language of the bees." As "language" has no essence, there is no compelling reason for not using it in a sense different from the one that Simpson evidently favors. Nonetheless, the criticism is just insofar as such usage might be misleading, tending to make the audience wrongly infer that the communication of bees has more in common with that of men than it really does. Thus, in spite of being in certain ways conventional, definitions can in fact be false, meaningless, or inappropriate. That definitions are not binding does not imply that anything goes.

In understanding how a definition can be *false,* it helps to draw a distinction between *stipulative definition* on the one hand, and *lexical definition* on the other. A "stipulative definition" is a proposal for the use of a term: "Let *x* designate *y*." Like the statement "Go away," it does not say anything true or false. On the other hand, a "lexical definition" asserts that a term is used in a given way: "People use the word *x* to designate *y*." One certainly may err in describing usage, and such lexical definitions as form the substance of dictionaries may indeed be incorrect. Another type of falsehood in definition would be an allegation that a definition is complete, in the sense of supplying all the defining properties, when it really is not. But we would not consider an incomplete definition false unless it really claimed to provide all of the defining properties. Thus, there is no falsehood involved when "*Salmo salar*" is defined as "a kind of fish."

We may encounter quite a variety of definitions that might be labeled as *meaningless.* They may be cast in terms of words which the audience does not know. Or they may be circular: "Homology is the opposite of non-homology." Or the term might refer to a nonsensical concept, such as a round square.

It is not always easy to decide whether a definition is *appropriate* or not, for it all depends on how we want to use the term. A chemist would define "sugar" in one way when teaching a chemistry course and in another for his young child. Much semantic dispute in the sciences arises from efforts to use a definition which is appropriate for one field, but not for another. The polemical literature on "homology" largely deals, not with how we should convey our thoughts about evolution, but rather with whether we should discuss it at all.

Let us now consider the terms "homology" and "analogy" to see how some of the definitions thus far proposed might be defective in such respects. For several reasons it helps to concentrate on the older literature. Often those who first deal with a concept have given it more thought than their successors, and frequently they did remarkably well. On the other hand, their mistakes have sometimes endured far beyond the time when correction became desirable.

The terms under consideration are in part the contribution of Richard Owen, one of the leading nineteenth century anatomists. He is generally credited with being the first to define "homology" and "analogy" in something close to their modern usage. On the other hand, as Boyden (1943) points out, both words had been used in somewhat different senses by earlier workers. Aristotle used "analogy," and both occurred quite often in the French and German literature for some time before Owen redefined them. Thus Isidore Geoffroy Saint-Hilaire (1832–1837, p. 59) said that homologies are the resemblances between the various parts of the same animal, analogies the resemblances between the equivalent parts of different animals. His father, Étienne Geoffroy Saint-Hilaire (1818, 1830) had used the term "analogues" for what would now be called "homologues" and Owen (1832) followed this usage in his classical monograph on the pearly *Nautilus*. Later Owen (1846) claimed to have done no more than render the older terms more explicit.

This older literature embodied concepts and habits of thought which now strike us as very odd. Our intellectual predecessors were given to a kind of thought process, analogizing, which was the drawing of far-fetched comparisons. We still do this, but to a much more limited extent, and we are usually quite aware that such "analogies" as the one between organisms and states are perhaps at best a pleasing way to embellish our language. Not so, many of the traditional naturalists. They believed that analogies had symbolic or mystical significance. Étienne Geoffroy Saint-Hilaire (1818, p. 371) hesitated to embrace numerology, but he pointed out that there are seven basic tones, seven vowels, etc., and he developed a theory that the vertebrate skull consists of seven vertebrae. Oken (1847) was even more extreme, finding threes, fives, and sevens everywhere. For MacLeay (1825) and his followers, the plan of Creation was a taxonomic scheme of circles within circles, each circle being made up of five smaller ones. "Affinity" was the relationship which linked, through point-by-point agreement in structure, the various forms and determined their position in the system. "Analogy" was a "correspondence between certain parts of the organization of two animals which differ in their basic structure." These analogies revealed the supernatural order. For example, Kirby and Spence (1826) thought that there was deep mystical significance to the analogy between predatory insects, and predatory fish, birds, mammals, and human beings. Swainson (1835) found circles of five in biogeography as well.

Reconsidering these older notions, we can see that although the nomenclature had not been stabilized, and although much metaphysics entered in, nonetheless some important distinctions were being drawn. *Corresponding parts* were being compared (Ghiselin, 1969a). The correspondences were of various sorts. A single organism's hand and foot have one relation, its left and right feet have another, and feet in different organisms yet additional ones. Owen did a real service to anatomy when he classified the different sorts of correspondence and provided terms for them. However, he made some rather unfortunate choices when he formulated his definitions.

Owen (1848, p. 2) referred to the homologous parts as "corresponding," but he missed the opportunity for treating both homology and analogy as kinds of correspondence. The trouble arose from his manner of bringing in function. He gives (Owen, 1843, p. 379) the following definition of "homolog": "The same organ in different animals under every variety of form and function." "Analog" means (Owen, 1843, p. 374) "A part or organ in one animal which has the same function as another part or organ in a different animal." He realized that parts which can readily be mistaken for homologs often have the same function, but erroneously concluded that no more is involved. It does not follow that since community of function is not a defining property of "homolog" it is a defining property of "analog." Compare defining "male" as "a member of the sperm-forming sex, irrespective of whether he is a child," and "female" as "a child." If Owen's terminology is followed, things can be both analogous and homologous at the same time (Owen, 1848, p. 7). They can also be one but not the other. This makes sense: some, but not all children are male. If we desire a good contrast term for "homology" we run into problems here. Other solutions lead to serious paradoxes. For instance, if we define "analogy" as "nonhomology" we obviously are not saying enough, for ears are then by definition analogous to furniture. Consider how "female" means more than "not male." Even defining "analog" as a "nonhomologous feature" will not do, for ears are then analogous to feet (see Bock, 1963).

Owen could have used his terms in conformity with the definitions he enunciated, for he was just as free to stipulate as anybody. But in examining his works, one can find paradoxes and inconsistencies, implying that he had one definition of "analog" for the glossary, another for the text. Our problem of interpreting him is difficult, however, because he often uses the term "analogy" in a nontechnical sense. Indeed, in the work where he first defines "homolog" and "analog" (Owen, 1843), the former term is rarely used, not occurring until page 141. Both senses of "analogy" occur, however, the technical one mainly later. Evidently he changed terminology. Nonetheless, Owen's very ambiguity suggests that he was not aware of the difficulties. He often makes statements like, "The alimentary canal of the Holothuria is closely analogous to that of the Echinus . . . " (Owen, 1855, p. 209). Most of these can probably be dismissed as either ambiguous or nontechnical usage. But in some instances there are serious paradoxes. For example, Owen (1843, p. 193) refers to the "analogues" in vertebrates and insects: thorax and anterior and posterior limbs. The thoraces, however, are just "similar." They do not have the same function. We could affirm that the first and the last pair of insect limbs have the same function as the anterior and posterior ones in vertebrates. But this produces a dilemma when Owen says that the middle pair in hexapods have "no analogue in the vertebrate series." Clearly, a locomotory function characterizes all these limbs. Again, Owen (1855, p. 278) notes that certain skeletal elements in boring bivalves are comparable to ones in stalked barnacles had been used as evidence for a relationship between the two groups; but he remarks that the "relation is one of analogy only." It seems unlikely that the two have the same function; they do, however, occupy equivalent positions. There might be functional or physiological reasons for such parts resembling each other, but they do not have the same function, or physiological role. The parts correspond in the same way to an abstract morphological scheme.

Consider, too, how Owen *does not* use "analogy." If we accepted his definition, we should have to say that a mushroom's (protective) poison is analogous to a clam's (protective) shell. Neither Owen, nor, so far as I know, anybody else, uses analogy in this fashion. A mushroom's poison is analogous to a newt's poison, and a clam's shell is analogous to an ostracod's shell.

It seems, therefore, that Owen's definition of "analogy" is not strictly lexical. If, as I have suggested (Ghiselin, 1969a), "analogy" is treated as a kind of correspondence, one contrasted with "homology," we come fairly closely to the actual intension of the term. Owen was distinguishing between what merely look like homologs and the real thing. His analogy concept was morphological, not physiological. He could not say what the difference between the two was, at least in any causal sense that will satisfy us today, because theory had not advanced enough. At the present time we use these terms when we seek to distinguish between organisms in a phylogenetic sequence (homologies) and the more superficial kind that have to be explained in some other way (analogies). Function does enter in, because the misleading, superficial resemblances are largely the result of evolutionary convergence in which dissimilar ancestors have given rise to descendants with comparable features. On the other hand, as Simpson (1961) points out, some parts resemble one another without sharing any function and for reasons other than convergence. The results of mimicry and coincidence are familiar examples. It is legitimate, but not yet a widely adopted practice, to use the term "analogy" here too.

Owen's definitions may also be criticized as meaningless. He says that homologues are "the same" organ, but gives no explicit criterion for telling whether a thing is the same as another or not (see also Simpson, 1961). In modern terminology, we do have such a criterion: homologues have the same ancestral precursor (Ghiselin, 1966b). The other way around this problem is to compare the homologues to an equivalent member in an abstract body plan, but in this case "the same" means something only when the plan itself is specified. Owen himself, however, could treat body plans as more than just abstractions, for they existed in the mind of God, as essences. He explains his views on the significance of homological anatomy as follows:

> The "Bedeutung," or signification of a part in an animal body, may be explained as the essential nature of such part—as being that essentiality which it retains under every modification of size and form, and for whatever office such modifications may adapt it. I have used here the word "Nature" in the sense of the German "Bedeutung," as signifying that essential character of a part which belongs to it in its relation to a predetermined pattern, answering to the "idea" of the Archetypal World in the Platonic cosmogony, which archetype or primal pattern is the basis supporting all the modifications of such part for specific powers and actions in all animals possessing it, and to which archetypal form we come, in the course of our comparison of those modifications finally to reduce their subject [Owen, 1849, p. 2].

But Owen maintained that one may deal with homology without reference to archetypes. This led him to distinguish between general and special homology, respectively, with and without the archetype. One could simply identify the corresponding parts in two animals, and forget the body plan. Yet whenever two entities are compared in this way, they are tacitly treated as members of a single class. And when we decide that there are relations between such entities, they must be relations of one sort or another, so an abstraction or its empirical equivalent is still present. To say that two things correspond, and nothing more, is like saying that they are near to one another, without reference to whether it is houses or galaxies being discussed.

The archetypes fitted in quite nicely with Owen's outlook, but they are harder to reconcile with our own. He belonged to a tradition that had asked, "If the Creator is one, why not the creation?" Unity was sought at the expense of diversity. The ideal classification scheme was conceived as embracing series from lower to higher organisms rather than as proceeding in various directions. Thus, although Owen did not posit a single continuum, his classifications retained much of the character of a "ladder" or *scala naturae* rather than that of a "phylogenetic tree" (see Hodos & Campbell, 1969).

This search for unity is also reflected in Owen's treatment of serial homology and related notions. It is a brute fact that lots of organisms are built up of repeated units having similar, if not identical, arrangements of their components.

Finding correspondences between, say, fingers and toes, or between organs on the left and right side of the body, is one kind of homologizing. Among these, Owen named only one kind. This was "serial homology," the correspondence between equivalent units in metameres–for example, between humerus and femur. He left the others unnamed, and did not provide a general term for all such corresponding elements in the same organism. Indeed, there would seem to be no term for this larger class, although Gegenbaur (1878, pp. 64–65) used "general homology" in a sense close to it and Smith (1967) uses "serial homology" and "homonomy" this way. Owen's view of serial homology was closely linked to his advocacy of the vertebral theory of the skull. He sought to demonstrate that the whole body is made up of smaller, repeated units for which there is a single archetype. Owen (1848) naturally explained these repetitions in idealistic terms. It pleased the Creator to use a single plan for the various parts of the body. Today serial homology provides a serious difficulty for anyone trying to provide a definition, because the precise theoretical significance of metamerism and allied phenomena is not entirely clear. The repeated parts are formed in much the same way in their embryological development and they have the same "plan," but we are not sure why. At any rate, more is involved than community of evolutionary descent. The correspondence nonetheless seems to be something fundamental in the organization of the animals, not superficial like analogy; hence, it is not inconsistent to continue calling it "homology."

A way in which Owen's approach seems inappropriate has to do with the relationships between form and function. He rejected what he called a "teleological" view of homology–that the differences and similarities between organisms could be explained on the basis of function alone. Yet he viewed the changes that have taken place among organisms through time as purposeful in a very old-fashioned sense, even as he begrudgingly accepted evolution. For example, Owen (1860) writes:

> One other conclusion may be drawn from a general retrospect of the mutations in the forms of the fishes at different epochs of the earth's history,–viz., that those species, such as the nutritious cod, the savoury herring, the rich-flavoured salmon, and succulent turbot, have greatly predominated at the period immediately preceding and accompanying the advent of man; and that they have superseded species which, to judge by the bony Garpikes (*Lepidosteus*), were much less fitted to afford mankind a sapid and wholesome food [p. 151].

Paradoxically, Owen's mixed attitude toward teleology in anatomy seems to have been largely responsible for his definition of "analogy." He conceived of the body plan as purely ideal, something by no means utilitarian in nature. Departures from the basic scheme would, however, be explicable in functional or adaptive terms. God would have one and only one reason for bringing superficial resemblances into existence: to provide for the needs of the organisms (see especially Owen, 1854). Such possibilities as historical accident and mimicry do not occur to one of his mentality; the problem is still with us.

With the Darwinian revolution, what had been called the Creation became many, not one. The Creator became an unnecessary hypothesis, and with Him went the teleological conception of the universe. Adaptation became an immediate and individualistic matter, the consequence of reproductive competition between members of the same species; fish could never have evolved for our benefit. Variation was now important, because the differences between individuals determine survival and reproductive success. Diversity took on an increasing significance in classification systems, for in addition to any progression from "lower to higher" creatures, an adaptive radiation was evident. Different groups had come to fill different ecological niches. The ladder was replaced by the tree.

Many of the problems surrounding homology were immediately resolved by the new evolutionary outlook. The term "homology" itself is now widely understood to mean "evolutionary homology," and such correspondences as those between serial homologs are considered less important. Common descent explains why it is possible to construct body plans and homologize from one group of organisms to another. On this point I cannot agree with de Beer (1971, p. 16), who says, "It is now clear that the pride with which it was assumed that the inheritance of homologous structures from a common ancestor explained homology was misplaced; for such inheritance cannot be ascribed to identity of genes." Such considerations are beside the point, for what matters is continuity in an historical sequence. Entities are homologous however different they become, and they can evolve new relationships to the chromosomes just as well as they can alter their structure or function. The problem is analogous to that of the changing connections of appendages to metameres (see Goodrich, 1910).

Homology has thus been redefined in historical terms, but resistance to this change still persists (see Naef, 1926; Boyden, 1943, 1947, 1969; Remane, 1952; Withers, 1964; and Jardine, 1967, who oppose; and Haas and Simpson, 1946; Simpson, 1961; Bock, 1963; and Ghiselin, 1969a, b, 1972, who favor the evolutionary definition). The opposition frequently maintains that although comparative anatomy can be interpreted in evolutionary terms, the Darwinian revolution did not alter its methodology. We must grant that the old methods of systematics and the old criteria of homology were never totally abandoned. And some workers still use techniques that are strictly nonevolutionary. But this does not mean that nothing has changed, as can readily be seen by an examination of Owen's systematics. He tried to arrange his materials so much as he could in the form of a *scala naturae.* This led him to very forced treatments of certain groups as transitional forms: barnacles between annelids and parasitic crustaceans, ichthyosaurs between fishes and reptiles, turtles between reptiles and birds, and so forth. Such arrangements made a good formal scheme, but poor history. Likewise, Owen (1882) was among the last to champion the unity of type for cephalopods and vertebrates. He homologized the mouth of the former with the hyphophysis and pineal gland of the latter; albeit formally consistent the scheme was not up-to-date biology.

The new comparative anatomy developed techniques of historical reconstruction that had no precedent in the old. Fundamental to the new approaches was the fact that natural selection is itself a powerful tool in the investigation of evolutionary history. It tells us that certain hypotheses about phylogeny are preferable to others. The best example has to do with the interpretation of vestigial organs. Owen (1849, p. 86) wrote, "For the Divine mind which planned the Archetype also foreknew all its modifications." Hence there was nothing wrong with fish' evolving for man's future use, or organs evolving for the sake of future generations but having no present function. Natural selection, unlike God, lacks foresight and cannot prepare organisms for life in some future environment. Useless features, therefore, are better seen as remnants than anticipations. It would help, particularly in the study of behavioral evolution, if such modern approaches were better known (see also Ghiselin, 1972). The available reviews (e.g., Wickler, 1961; Atz, 1970; Campbell & Hodos, 1970) pay little attention to them, although there are exceptions (e.g., Hinde & Tinbergen, 1958).

One can easily see that concepts like similarity and homology are readily misunderstood. The best way to avoid logical mistakes when using these concepts is to analyze them and be aware of the common fallacies. In this connection we should briefly consider some of the pitfalls which one ought to avoid. In the first place, we need to distinguish, very carefully, between similarity, on the one hand, and correspondence, on the other. In the following diagram, entities composed of parts designated by letters are labeled 1, 2, 3, and 4 and are arranged according to "how similar" they are:

1 *a–b–c*
2 *a–b–d*
3 *a–e–d*
4 *f–e–d*

Here, 1 is similar to 2, 2 is similar to 3, and 3 is similar to 4. But this does not imply that 1 is similar to 4. In 1 and 4, however, the parts designated with letters do correspond to each other, and they correspond just as much as do the parts of 1 and 2. There is no necessary connection between correspondence and similarity.

A second source of confusion arises through uncertainty as to whether it is wholes or parts that are being compared. Is it similar (or corresponding) organisms, or similar (or corresponding) organs? (1 and 2 or a and f in the diagram above.) Analogs are often in fact the corresponding parts of similar organisms; but they are treated all too often as if they had to be similar themselves, especially similar in function.

A third mistake is to infer that evolutionary homologs must be parts of different organisms. But two parts can be derived from a single one in an earlier stage of evolutionary history. For example, many enzymes are thought to have arisen by gene duplication. Two copies, as it were, of a single enzyme were produced, and these gradually diverged. The corresponding amino acids which make up the modified proteins would be homologous.

Fourthly, misconceptions arise through a tacit assumption that only tangible objects (or parts in a restricted sense) can be homologized. Many have reasoned that behavioral entities, being activities rather than organs, lack a morphology or structure. Yet they are often made up of sequences of comparable events, and the elements of these sequences can be treated much like the parts of bones in a skeleton or amino acids in a protein molecule.

Finally, we have a tendency to confound genetics with genealogy. Behavioral materials can be homologized irrespective of whether the behavior itself is acquired or innate. Indeed, comparative linguistics has a most elaborate system of correspondences which do not differ in principle from those of comparative anatomy. There is a far greater potential for unity between the behavioral and the biological sciences than some would like to admit.

CONCLUSIONS: SOME SUGGESTED DEFINITIONS

Proposals for improving the scientific vocabulary tend to be somewhat immodest, and perhaps it would be better to leave well enough alone. Yet it does seem that there is room for improvement. We should keep in mind the need to keep our definitions lexical, appropriate, and meaningful, and I have tried to do so. I have attempted to give definitions which are reasonably complete, and this necessitates rather complicated "formal" definitions. One is likely to be taken to task for laziness if one leaves these out, for obscurity if one includes them. Such expressions are appropriate for some uses but not for others. They are useful in philosophical analysis, and whenever it is desirable to make sure that assertions are rigorous and unequivocal. As means of explaining terms to beginners, they have little value: here the appropriate literary form is an essay, one rich in examples and not necessarily terse.

A.1. *Homology.* "Entities are homologous when they are elements of members of a group of precursors and derivatives having a common source, and when these entities can, in principle, be traced back through the derivational series to the same element of a precursor from which the members derive, the sense in which they are homologous being stipulated so as to avoid ambiguity."

B.1. *Analogy.* "Entities are analogous when they are nonhomologous corresponding elements of a class of more inclusive entities which share properties suggestive of a genuine homology between those elements."

Remarks on A.1. and B.1. Briefly and not very accurately paraphrasing, homologues are parts that arise from the same source; analogues are parts which appear to be homologues but really are not. The terms are contraries: entities cannot be homologous and analogous *in the same sense.* "Homology" is a strictly empirical concept, whereas "analogy" is part of the language used to distinguish between fact and theory.

A.2. *Kinds of homology.* Here, these are classified according to the kind of precursor. Note that the list is not exhaustive, and that the terms may be applied at several levels of integration. Thus, one may homologize the parts of faunas as well as those of organisms.

A.2.a. *Evolutionary homology.* "Entities are homologous when they can be traced back to the same common ancestral precursor." This definition is roughly equivalent to that provided elsewhere by Ghiselin 1966b except that it of course presupposes the more general concept given above under "A.1." Examples: forearm of a bat and a bird (stipulated as forearm, but not as wing); bison faunas of North America and Eurasia; *esse* in Latin and *être* in French.

A.2.b. *Iterative homology.* "Entities are iteratively homologous when they can be traced back to the same element of a precursor which causes the repetition of parts in a larger whole." Note that reference to body plans has been avoided. The term, to my knowledge, is new. Under it fall such relations as serial homology, but others should be recognized as well. Cytologists refer to "homologous chromosomes"—the corresponding pairs in a diploid set. In the behavioral sciences, "iterative homology" should occur wherever groups of behavioral phenomena are repeated, especially with variations. Thus, one might wish to treat the genitive singulars in the first and second declension of Latin as iterative homologs. The following are terms for some kinds of iterative homology recognized by anatomists.

A.2.b.i. "*Serial homology* is the relation between the iterative homologues of parts which are repeated in linear sequence." Example: femur and humerus.

A.2.b.ii. "*Antimeric homology* is the relation between the iterative homologues of parts that are repeated on the opposite sides of the body." Example: right and left humerus.

A.2.b.iii. "*Sexual homology* is the relation between the iterative homologues of reproductive parts of organisms in species in which a single kind of reproductive apparatus is repeated with variations in individuals having different reproductive roles." The concept is very old, the term, I think, my own (Ghiselin, 1969b), the definition new. Example: penis and clitoris.

B.2. *Kinds of analogy.* I like to distinguish the following, but again, the list is not exhaustive. Note that for none of them is a common function the sole defining property, although it helps.

B.2.a. *Convergent analogy.* "Entities are convergent analogues when the properties which make them analogous are present as the consequence of convergent evolution." In many instances this term can replace homoplasy, homomorphy, and the like. Examples: wing of bird and wing of insect; "marsupial wolf" and true wolf as elements of their respective faunas.

B.2.b. *Parallel analogy.* "Entities are parallel analogues when the properties which make them analogous are present as the consequence of parallel evolution." The difference between parallel and convergent evolution is a matter of degree, but still an important one. The term replaces the very confusing "homoiology." Example: wing of bird and bat (as wing but not as anterior appendage).

B.2.c. *Iconic analogy.* "Entities are iconic analogues when the properties that make them analogues are present as the consequence of selection having favored a resemblance (as such) between the entities of which the elements are analogized." An "icon" is an image. Iconic analogy covers the results of mimicry and camouflage. Example: the leaf of a tree and the leaflike wing of an insect camouflaged by its resemblance to part of a tree.

B.2.d. *Coincidental analogy.* "Entities are coincidental analogues when the properties which make them analogous are present as the consequence of nothing more than fortuitous circumstances." This term covers the "chance similarity" of Simpson (1961). Example: leg of man, root of a mandrake.

I shall conclude with a few remarks on how much terminology is really necessary. "Homology" and "analogy" as defined above, admitting as they do of subclasses, are all that comparative anatomists really need. In ordinary discourse, among scholars in the field, the terms are used in stating whether parts are homologous or analogous, either in asking the question or in giving the answer. In such discourse, we really are not fundamentally interested in whether features have the same use, and we do not even need a term for this. "Analogous," in the sense of having a common function, is not used in the ordinary discourse of biologists; it occurs virtually only, if not exclusively, in discussions of semantics alone. So it is with a host of other words: homoplasy, homomorphy, homogeny, homoilogy, isology, etc. Their value is largely pedantic. A scientist's language should be adapted to his needs. If an anatomist or other comparative biologist can say all he wants, and say it well, with minimal effort, it does not pay to burden his memory with an inflated lexicon.

ACKNOWLEDGEMENTS

Preparation of this chapter was supported by Grant GS 33491 from the National Science Foundation. For comments, not always in agreement with my views, I thank William Hodos and Bobb Schaeffer. For advise on style and grammar I thank Brewster Ghiselin.

REFERENCES

Atz, J. W. In L. R. Aronson, E. Tobach, D. S. Lehrman, & J. S. Rosenblatt (Eds.), *Development and evolution of behavior: Essays in memory of T. C. Schneirla.* San Francisco: Freeman, 1970. Pp. 53–74.

Bock, W. J. *American Naturalist,* 1963, **97**, 265–285.

Boyden, A. *Quarterly Reviews in Biology,* 1943, **18**, 228–241.

Boyden, A. *American Midland Naturalist,* 1947, **37**, 648–669.

Boyden, A. *Science,* 1969, **164**, 455–456.

Campbell, C. B. G., & Hodos, W. *Brain, Behavior & Evolution,* 1970, **3**, 353–367.

de Beer, G. *Homology: An unsolved problem.* London: Oxford University Press, 1971.

Gegenbaur, C. *Elements of comparative anatomy.* (Translated by F. J. Bell) London: Macmillan, 1878.

Geoffroy Saint-Hillaire, É. *Philosophie anatomique.* Paris: Méquignon-Marvis, 1818.

Geoffroy Saint-Hilaire, É. *Principes de philosophie zoologique discutés en Mars 1830, au sein de l'Academie Royale des Sciences.* Paris: Pichon et Dider, 1830.

Geoffroy Saint-Hilare, I. *Histoire générale et particulière des anomalies de l'organisation chez l'homme et les animaux, ouvrage comprenant des recherches sur les caractères, la classification, l'influence physiologique et pathologique les rapports généraux, les lois et les causes des monstruosités, des variétés et vices de conformation, ou traité de tératologie.* Paris: J.–B. Balliere, 1832–1837. 3 vols and Atlas. **1**, 943 pp.

Ghiselin, M. T. *Philosophy of Science*, 1966, **33**, 147–153. (a)

Ghiselin, M. T. *Systematic Zoology,* 1966, **15,** 127–130. (b)

Ghiselin, M. T. *Systematic Zoology,* 1969, **18,** 148–149. (a)

Ghiselin, M. T. *The triumph of the Darwinian method.* Berkeley, California: University of California Press, 1969. (b)

Ghiselin, M. T. In T. J. M. Schopf (Ed.), *Models in paleobiology.* San Francisco: Freeman, 1972. Pp. 130–145.

Goodrich, E. W. *Quarterly Journal of Microscopical Science*, 1910, **59**, 227–248.

Haas, O., & Simpson, G. G. *Proceedings of the American Philosophical Society,* 1946, **90,** 319–347.

Hinde, R. A., & Tinbergen, N. In A. Roe & G. G. Simpson (Eds.), *Behavior and evolution.* New Haven: Yale University Press, 1958, Pp. 251–268.

Hodos, W., & Campbell, C. B. G. *Psychological Reviews*, 1969, **76**, 337–350.

Jardine, N. *British Journal of Philosophy Science*, 1967, **18**, 125–139.

MacLeay, W. S. *Transactions of the Linnean Society,* 1825, **14,** 46–68.

Naef, A. *Biol. Zentralblatt,* 1926, **46,** 405–427.

Oken, L. *Elements of physiophilosophy.* London: Ray Society, 1847.

Owen, R. *Memoir on the pearly nautilus* (Nautilus pompilius, Linn.) *with illustrations of its external form and internal structure.* London: Royal College of Surgeons, 1832.

Owen, R. Lectures on the comparative anatomy and physiology of the invertebrate animals, delivered at the Royal College of Surgeons, in 1843. London: Longmans, Brown, Green, and Longmans, 1843.

Owen, R. *Edinburgh Phil. Mag. J.*, 1846, **28,** 525–527.

Owen, R. *On the archetype and homologies of the vertebrate skeleton.* London: Richard and John E. Taylor, 1848.

Owen, R. *On the nature of limbs.* London: John van Voorst, 1849.

Owen, R. *The principal forms of the skeleton and teeth.* Philadelphia: Blanchard and Lea, 1854.

Owen, R. *Lectures on the comparative anatomy and physiology of the invertebrate animals, delivered at the Royal College of Surgeons.* London: Longmans, Brown, Green, and Longmans, 1855.

Owen, R. *Palaeontology or a systematic survey of extinct animals and their geological relations.* Edinburgh: Adam and Charles Black, 1860.

Owen, R. *Journal of the Linnean Society (Zoology),* 1882, **16,** 131–149.

Remane, A. *Die grundlagen des natürlichen systems, der vergleichenden anatomie und der phylogenetik.* Leipzig: Akademische Verlagsgesellschaft Geest and Portig, 1952.

Robinson, R. *Definition.* Oxford: Oxford University Press, 1954.

Simpson, G. G. *Principles of animal taxonomy.* New York: Columbia University Press, 1961.

Simpson, G. G. *Science,* 1966, **152,** 472–478.

Smith, H. M. *Systematic Zoology,* 1967, **16,** 101–102.

Swainson, W. *A treatise on the geography and classification of animals.* London: Longman, 1835.

Wickler, W. *Forschritte Zoologie*, 1961, **13**, 303–365.

Withers, R. F. J. In J. R. Gregg & F. T. C. Harris (Eds.), *Form and strategy in science: Studies dedicated to Joseph Henry Woodger on the occasion of his seventieth birthday.* Dordrecht: Reidl, 1964. Pp. 378–394.

Wittgenstein, L. *Tractatus logico-philosophicus.* London: Routledge & Kegan Paul, 1961.

10
Morphological Homology and the Nervous System

C. B. G. Campbell

Indiana University[1]

After an extended period of time in which little attention was paid to problems involved in the application of the concept of homology to studies of the evolution of the brain and spinal cord, Nieuwenhuys and Bodenheimer (1966) presented a discussion of the criteria for establishing central nervous system homologies as a part of a study on the diencephalon of the bony fish *Polypterus*. These authors presented a brief review of the subject and also their view as to which "criterion" of homology was most useful. This appeared at the same time as a series of papers by Ghiselin (1966a, b) on the fundamental basis and definition of the concept in general. In addition, dramatic findings by Karten and his co-workers (Karten, 1967, 1968, 1969; Karten & Hodos, 1970) have initiated an era in which the old concepts of the evolution of the forebrain are now seriously challenged as a result of new and different conclusions concerning homologies in the telencephalon of amniotes. The coincidence of these events has prompted a review of current thinking on the concept of homology and reexamination of the classes of data available for the recognition of homology in the central nervous system (Campbell & Hodos, 1970). A similar discussion has been presented by Northcutt (1970). Most of what follows below is taken from Campbell and Hodos (1970).

Students of the evolution of the nervous system, just as other evolutionary scientists, have only two sources of evidence on which to base their inferences. These are an examination of the fossil record and the comparative study of the nervous systems of living animals. Vertebrate paleontology is itself the study of the comparative anatomy of those portions of the bodies of vertebrate animals

[1] Current Address: Departments of Radiological Sciences and Anatomy, University of California, Irvine, California.

that are capable of fossilization. Comparative anatomy and its methodology are, therefore, at the very center of evolutionary study (along with modern genetics).

Similarly, the concept of homology is central to the methodology of comparative anatomy. If some portion of the nervous system in a series of different animals is being compared with the intention of inferring the course of change in that structure within a particular lineage through time, then it must be possible to recognize which structures are truly comparable in this sense. Only homologous structures should be compared for this purpose.

DEFINITIONS

The biological concept of homology stems from the British anatomist Richard Owen, who describes a homolog as "the same organ in different animals under every variety of form and function." Two major schools of thought have arisen around the concept of homology. The first school bases the concept of homology on structural correspondence and the second bases it on common ancestry. I shall not enter into a discussion here of the relative merits of these two approaches. I have discussed my views on this issue elsewhere (Campbell & Hodos, 1970). Suffice it to say that I am an adherant of the common ancestry school and the definitions of homology and related terms (homogeny, homoplasy, and analogy) that I accept are based on common ancestry. The definitions of these terms that I prefer are derived from Ghiselin (1966a,b), Bock (1969), Lankester (1870), and Simpson (1961). Although I have listed them previously (Campbell & Hodos, 1970) it is perhaps useful to list them again:

Homology. Structures and other entities are homologous when they can, in principle, be traced back through a genealogical series to a stipulated common ancestral precursor irrespective of morphological similarity.

Homogeny. Homologous structures that are also morphologically similar are homogenetic.

Homoplasy. Structures or entities that are morphologically similar but that cannot, in principle, be traced back to a stipulated common precursor are homoplastic. The concept of homoplasy encompasses:

Convergence. The appearance of similar characters separately in two or more lineages without a common ancestry pertinent to the similarity but involving adaptation to similar ecological conditions.

Parallelism. The appearance of similar characters separately in two or more lineages of common ancestry and on the basis of, or channeled by, characteristics of that ancestry.

Chance similarity. Resemblance in characteristics that appear in separate groups by independent causes but not involving adaptation to similar ecological conditions.

Analogy. Correspondence between structures or entities because of similarity in function whether or not they can be traced to a stipulated common precursor.

Ghiselin (1966a) emphasizes the importance of stipulating the basis of the homology and gives the example of the wings of eagles, which are homologous to those of hawks as "wings" and to those of bats as "derivatives of forelimbs." In simplistic discussions of the concepts of homology and analogy the same example is often given to illustrate homology in the first case (eagles versus hawks) and analogy in the second (eagles versus bats). An analogous relationship does exist in the second case but, as Ghiselin indicates, a homologous relationship that can be stipulated exists also. A similar example can be drawn from the central nervous system. The motor nucleus of cranial Nerve XII in the domestic cat is homologous to the motor nucleus of XII in the tiger as motor neurons innervating the skeletal muscle of the tongue and is homologous to the nucleus of cranial Nerve XI in the iguana as derivatives of the basal plate. The mere statement that Structure A is homologous to Structure B is not sufficient.

CRITERIA FOR THE RECOGNITION OF HOMOLOGY

The conception that the definition of homology should not be tied to morphological similarity has been discussed elsewhere (Campbell & Hodos, 1970; Ghiselin, 1966a,b; Smith, 1967). Similarity is, however, the most important basis for the recognition of homology. Convergence and parallelism, two forms of homoplasy, are very frequent sources of superficial similarity that can mislead the student of evolution. Every effort should be made to exclude these sources of confusion when evaluating evidence of homology. It should be recognized, however, that it is not always possible to do this.

An excellent discussion on criteria for homology is to be found in Simpson (1961). He points out that when examining characters held in common by two or more animals, the minuteness of resemblance and multiplicity of similarities are important criteria. As valuable as these criteria are, they do not totally eliminate the possibility of homoplastic similarity. It should also be remembered that groups of characters that are closely correlated on either a genetic or a functional basis should be considered as one item, not as independent items of evidence. Simpson (1961) observes that the evidence of fossil lineages converging backward in time toward a common ancestor is a direct criterion of homology. It should be remembered that the common ancestor may not be a single species but an ancestral genus, family, or other supraspecific group. The classic example of homology, that between mammalian ear ossicles and reptilian jawbones, was established by this sort of evidence. This criterion is not really useful for the internal anatomy of the central nervous system but is potentially useful for certain gross aspects of brain morphology.

Radinsky (1968, 1970, 1971), using latex endocranial casts, has shown that quite useful information on cortical sulcal patterns (and therefore such related aspects as somatic sensory cortical projection areas), relative development of occipital, temporal, parietal, and frontal lobes, etc., can be studied in fossil series. Radinsky was not concerned with establishing homologies; nevertheless, his work pointed up the potential of this improved method coupled with the more sophisticated knowledge available today. It is of course true that although the time dimension is added when a fossil series is compared, the problem of inferring ancestor–descendant relations remains.

Another criterion for the recognition of homology is ontogenetic similarity. The developmental patterns that result in the production of a particular organ, such as the brain, are inherited. Consequently, strong evidence of homology can be derived from the demonstration of a common embryological origin.

CLASSES OF DATA FOR RECOGNIZING HOMOLOGY

The central nervous system of vertebrates has its ontogenetic origins in an elongate flat plate of ectoderm that forms on the dorsal surface of the embryo. This plate rolls up to form an elongate tube and no matter in what group of animals it may be found, or how profoundly modified, traces of its tubelike nature always remain. Stem cells originally found surrounding the cavity of the tube give rise to the supporting cells of the central nervous system, the neuroglia, and to the major functional components, the neurons. In some animal groups, e.g., the amphibians, the bulk of the neurons formed remain more or less in proximity to the central canal, whereas in others considerable migration of neurons toward the periphery occurs.

Neurons have the tendency to occur in groups or populations of like cells with common functions that may be organized in a rather loose manner, as in the so-called "reticular formation," in more or less densely packed masses called "nuclei" or "nuclear groups," or in some special instances in successive layers or "cortical" formations. There are undoubtedly certain functional advantages and disadvantages to each of these patterns of organization.

The central nervous system functions as a communications and regulatory agency because the various populations of neurons receive inputs of information from other neurons or sensory receptors impinging on their cell bodies and cell processes, integrate this information, and transmit new data. In general, dendritic processes carry information toward the soma and axonal processes carry information away from the soma to other neurons or to effectors, e.g., muscles or glands. The axons of one or more population of neurons may run together in a common bundle or tract and be sufficiently conspicuous as to be visible on inspection either unstained or with the aid of tissue stains. Often axons of a common functional class with their origins in the same population of neurons or related populations travel among other axons in such a manner as to be nondifferentiable from them without demonstration by experimental methods.

Until very recently the bulk of neuroanatomical studies were performed either by examining normal brains grossly or by sectioning them into thin slices and straining them with dyes that stained either the cell somata or the cell processes (primarily the axons). Each of these methods alone gives a very incomplete, and in some ways misleading, picture of how the nervous system is organized. Some older experimental methods of limited usefulness, e.g., retrograde chromatolysis or the Marchi method for demonstrating the degenerating myelin sheaths of axons, have been applicable in certain instances. The Golgi method makes visible cell somata and their processes simultaneously; however, only a small and unpredictable number of neurons are stained in any one section and the method is not useful for neurons the axons of which project any great distance from the cell body.

Because the routine staining methods emphasize either the groups of neuron cell bodies or the groups of cell processes it has been common to talk about nuclear groups and "fiber connections" as if they are a series of telephone exchanges with interconnecting cables. I believe that this has had an unfortunate effect on some of those workers who have concerned themselves with the kinds of data that may be considered useful in drawing homologies in the central nervous system.

What kinds of data have been considered useful in attempts to infer homology? Nieuwenhuys and Bodenheimer (1966) point out that historically three kinds of data have been used to determine homologies in the central nervous system: (1) ventricular sulci or grooves that mark either the middle of areas of cell proliferation or the boundaries of functional areas, depending on whose interpretation you accept (Bergquist, 1932; Herrick, 1933; Källén, 1951; Kuhlenbeck, 1929); (2) the topographic and/or topologic position of nuclear groups; and (3) fiber tracts or connections between nuclear groups. (Nieuwenhuys and Bodenheimer's concern was primarily with the diencephalon; consequently their discussion was limited to ventricular sulci. Superficial sulci can be used for similar purposes.) A number of workers have advocated one or another of the three classes of data as most useful for establishing homologies, whereas Herrick(1933) considers that all three methods may lead to the same conclusions. Most workers, including Nieuwenhuys and Bodenheimer (1966), have not considered fiber connections to be particularly useful.

Campbell and Hodos (1970) suggest that it is not realistic to expect any single class of data to be universally useful, and that as many lines of evidence as can be found applicable to a particular analysis should be used. For example, the ventricular sulci that Kuhlenbeck (1929) advocated as the most important class of data for determining homologies in the diencephalon of anamniotes were found by Nieuwenhuys and Bodenheimer (1966) to be highly variable in their presence, distinctness, and position within the genus *Polypterus* and between

Polypterus and other genera of bony fishes. This does not imply that the ventricular sulci may not be quite consistent in these characteristics and therefore useful for drawing homologies in the diencephalon of urodeles, as Herrick has thought, or possibly even all amphibians. An extensive comparative study would have to be done to determine this. Sulci, ventricular or superficial, are to a degree limited in usefulness by their nature. That is, they are confined to surfaces and there are large areas of the nervous system in which they are not present.

Another class of data is the character and position of nuclear groups. Slices of nervous tissue are stained with basic dyes, such as thionin or Cresyl Violet, which stain essentially only the cell soma of neurons. The neuron cell processes, dendrites and axons, are not stained. On the bases of cell size, staining characteristics, packing density, the presence of cell-free gaps, etc., the central nervous system can be parcellated into different cell populations, each of which can be given a name or other designation. It is assumed that in the main these "nuclei" are populations of cells that share some unique functional role.

The morphology and topographic relations of these nuclear groups have most commonly been used to establish homologies between diverse groups of vertebrates. As mammalian brains are the most studied it has been not uncommon to apply mammalian neuroanatomical names to nuclear groups in nonmammalian vertebrates if they resemble the mammalian populations in one or more characteristics, e.g., configuration, staining density, general position, etc. Herrick (1948, p. 8) discusses the rationale for this practice and it is difficult to argue with his point that in the early stages of inquiry it is necessary to apply descriptive names to visible structures before their relationships are known. In spite of Herrick's caution that "homologies implied by similar names are rarely exact . . . " it has been easy to be lulled into the belief that a great deal is known about the organization and evolution of nonmammalian brains merely because of those familiar names. Modern experimental studies, by showing the true nature of many of these nuclei, have been revealing just how poor is current understanding. Cytoarchitectural evidence is nevertheless useful in many cases, although it should not be considered the only valid evidence.

In the past, fiber connections were not considered to be particularly useful as a class of data bearing on homology. Nieuwenhuys and Bodenheimer (1966) have perhaps been the most recent to express this view. These negative reactions are, I believe, a consequence of several factors. First, there has been a tendency to make an artificial dichotomy between nuclear groups and their fiber connections. Fiber connections most readily revealed by nonexperimental methods are those involving large bundles or tracts, often comprised of axons from more than one neuronal population and often destined for more than one terminus, e.g., the so-called "pyramidal system." This pathway is composed of axons of neurons in both the motor and the sensory areas of the cerebrum and the axons comprising this fiber system project to many different structures in the brainstem and spinal cord. The sites receiving pyramidal input vary both qualitatively and quantitatively when comparisons are made between the various orders of extant mammals using modern experimental techniques.

Nieuwenhuys and Bodenheimer (1966) argue that the use of fiber connections for drawing homologies involves a circularity of reasoning: defining fiber tracts on the basis of the cell masses they interconnect and then using the fiber bundle as the basis for homologizing cell masses. They further state that if a fiber tract is defined by a characteristic topographical position of one segment in the tract, then it cannot be certain whether tracts with the same topographical position at that segment connect nuclei with any similarity of morphological position. I believe that both of these arguments stem from thinking about fiber connections as if they were something other than the cell processes of neuronal populations (i.e., the telephone exchanges connected by cables view), and acceptance of the limitations of older methods of tracing fiber connections.

Since the mid-1950s, ever more sophisticated methods of tracing central nervous system circuitry have been devised, starting with the experimental silver impregnation methods for degenerating axoplasm introduced by Nauta and Gygax (1954). More recently, autoradiographic and other new tracing methods have been introduced that promise to reveal the axonal projections of deep-lying nuclear groups without the disadvantages of electrolytic lesion-making techniques usually used when applying the Nauta methods. The precise site of termination of an axonal projection can be determined on the postsynaptic cell and its processes using the appearance of certain characteristics of degenerating axon terminals as seen in the electron microscope. These new methods, combined with some of the older ones, make it possible to know the location, configuration, and input–output associations of populations of neurons. The circuitry of whole subsystems within the brain and spinal cord can be determined. With added physiological and behavioral information about the function of the subsystems, there is possible potentially a more real understanding of the organization of nervous systems than was possible in the first half of the twentieth century.

The central nervous system owes its unique properties and functions to the fact that its component parts are specialized populations of cells that communicate with other such cells and effector organs via elongate extensions of their own protoplasm. These associations of the neuron populations are certainly as important in characterizing them as the topographic location, staining characteristics, etc., of their cell somata. Indeed, there are some nuclear groups the functional role of which in the nervous system could not be legitimately speculated on until modern experimental studies of their fiber connections revealed that they belonged to some particular subsystem, e.g., nucleus rotundus thalami of birds and reptiles as part of the ascending visual pathway.

Recent work using modern degeneration techniques has been revealing the projection of the major ascending sensory systems to portions of the telencephalon, the dorsal ventricular ridge of reptiles, and its supposed homolog, the neostriatum of birds. These pathways are organized in a manner resembling those which project to the neocortex in mammals. Karten (1969) and others (Nauta & Karten, 1970; Hall & Ebner, 1970; Northcutt, 1970), on the basis of connectivity patterns as well as histochemical, physiological, and embryological evidence, have suggested that these cell masses, although not organized in layers, may be the homolog of neurons of portions of the neocortex of mammals. The cells of these areas, on the basis of cytoarchitecture, have been considered homologs of basal ganglia neurons of mammals.

Although more evidence is required before these suggested homologies can be considered to be established, the evidence currently available is highly suggestive. Questions of this dimension cannot be answered by utilizing any single class of data as the sole valid evidence of homology. As many lines of evidence as can be found to bear on a particular homology should be examined. New information, if valid, should be welcomed and old prejudices against fiber connections as data bearing on homology must be discarded.

REFERENCES

Bergquist, H. Zur Morphologie des Zwischenhirns bei niederen Wirbeltieren. *Acta Zoologica,* 1932, **13,** 57–304.

Bock, W. J. Discussion: the concept of homology. In J. M. Petras & C. R. Noback (Eds.), *Comparative and evolutionary aspects of the vertebrate central nervous system. Annals of the New York Academy of Science,* 1969, **167,** 71–73.

Campbell, C. B. G., & Hodos, W. The concept of homology and the evolution of the nervous system. *Brain, Behavior, & Evolution,* 1970, **3,** 353–367.

Ghiselin, M. T. An application of the theory of definitions to systematic principles. *Systematic Zoology,* 1966, **15,** 127–130. (a)

Ghiselin, M. T. On psychologism in the logic of taxonomic controversies. *Systematic Zoology,* 1966, **15,** 207–215. (b)

Hall, W. C., & Ebner, F. F. Thalamotelencephalic projections in the turtle *(Pseudemys scripta). Journal of Comparative Neurology,* 1970, **140,** 101–122.

Herrick, C. J. The amphibian forebrain. VI. *Necturus. Journal of Comparative Neurology,* 1933, **58,** 1–288.

Herrick, C. J. *The brain of the tiger salamander.* Chicago: University of Chicago Press, 1948.

Källén, B. Embryological studies on the nuclei and their homologization in the vertebrate forebrain. *Königlich Fysiografiska Sällskapet, Lund Handlingar. N. F.,* 1951, **62,** 34.

Karten, H. J. The organization of the ascending auditory pathway in the pigeon (*Columba livia*). I. Diencephalic projections of the inferior colliculus (nucleus mesencephali lateralis par dorsalis). *Brain Research,* 1967, **6,** 409–427.

Karten, H. J. The ascending pathway of the pigeon (*Columba livia*). II. Telencephalic projections of the nucleus ovoidalis thalami. *Brain Research,* 1968, **11,** 134–153.

Karten, H. J. The organization of the avian telencephalon and some speculations on the phylogeny of the amniote telencephalon. *Annals of the New York Academy of Science,* 1969, **167,** 164–179.

Karten, H. J., & Hodos, W. Telencephalic projections of the nucleus rotundus in the pigeon (*Columba livia*). *Journal of Comparative Neurology,* 1970, **140,** 35–52.

Kuhlenbeck, H. Über die Grundbestandteile des Zwischenhirnbauplans der Anamnier. *Morphology Journal,* 1929, **63,** 50–95.

Lankester, E. R. On the use of the term homology in modern zoology, and the distinction between homogenetic and homoplastic agreements. *Annals and Magazine of Natural History,* 1870, **6,** 34–43.

Nauta, W. J. H., & Gygax, P. A. Silver impregnation of degenerating axons in the central nervous system: A modified technic. *Stain Technology,* 1954, **29,** 91–93.

Nauta, W. J. H., & Karten, H. J. A general profile of the vertebrate brain, with sidelights on the ancestry of cerebral cortex. In F. O. Schmitt (Ed.), *The neurosciences: Second study program.* New York: Rockefeller University Press, 1970. Pp. 7–26.

Nieuwenhuys, R., & Bodenheimer, T. S. The diencephalon of the primitive bony fish *Polypterus,* in the light of the problem of homology. *Journal of Morphology,* 1966, **118,** 415–450.

Northcutt, R. G. *The telencephalon of the western painted turtle (Chrysemys picta belli).* Urbana, Ill.: University of Illinois Press, 1970.

Radinsky, L. B. Evolution of somatic sensory specialization in otter brains. *Journal of Comparative Neurology,* 1968, **134,** 495–506.

Radinsky, L. B. The fossil evidence of prosimian brain evolution. In C. R. Noback & W. Montagna (Eds.), *The primate brain: Advances in primatology,* Vol. 1. New York: Appleton-Century-Crofts, 1970. Pp. 209–224.

Radinsky, L. B. An example of parallelism in carnivore brain evolution. *Evolution,* 1971, **25,** 518–522.

Simpson, G. G. *Principles of animal taxonomy.* New York: Columbia University Press, 1961.

Smith, H. M. Biological similarities and homologies. *Systematic Zoology,* 1967, **16,** 101–102.

11

The Concept of Homology and the Evolution of Behavior

William Hodos

University of Maryland

The study of the evolution of behavior is the study of history. However, it is a study carried out with an enormous handicap. Unlike students of political, social, or economic history, evolutionary behaviorists have no archives, no old documents, reports, and letters to sift. Paleontologists at least have a record, albeit fragmentary, of events and organisms preserved in rock. For the behaviorist, however, the only record of the past is that which is written in the DNA molecule. The behavior of ancestors can only be known to the extent that it is expressed in the behavior of their descendents. Therefore the study of the historical development of behavior is carried out on living animals with particular groups of animals used as representatives of earlier stages in the development of other groups. Two main questions must be answered before such a study can be begun: first, which groups of animals shall be studied? Second, which characteristics of these animals shall be compared? I have discussed strategies and pitfalls in the selection of extant forms as representatives of ancestors elsewhere (Hodos & Campbell, 1969; Hodos, 1970, 1974) and shall not dwell on that matter here. Instead, I shall devote myself to the problems concerning the answer to the second question.

The goal of historical reconstruction of behavior is the description of progressive changes in a particular behavior through a series of contiguous populations of organisms representing specific ancestor–descendent relationships. In describing such a character gradient, the investigator must deal with equivalent features in the different populations (Mayr, 1970). A term that has been associated with such equivalence is "homology." However, the meaning of the term "homology" has been the subject of considerable controversy because various writers have used the term to indicate different types of equivalence (Campbell & Hodos, 1970; Northcutt, 1969; Källén, 1959; Campbell, Chapter

10 of this volume; Hailman, Chapter 14 of this volume). In its most simplified form, the contention between the opposing positions is whether homology should refer to characters that are equivalent because they are similar, irrespective of commonality of phyletic origin, or whether homology should refer to characters that are equivalent because they have a common phyletic origin, irrespective of their degree of similarity. Either use of the term describes relationships that are meaningful scientifically. However, because equivalence based on commonality of phyletic origin seemed to be more directly related to the problem of evolutionary reconstruction than structural similarity, Campbell and Hodos (1970) recommended the phyletic sense of the term when applying "homology" to comparative neuroanatomical material. They suggested the following definition of homology: "structures and other entities are homologous when they could, in principle, be traced back through a genealogical series to a stipulated common ancestral precursor irrespective of morphological similarity [p. 358]." The purpose of this chapter is to extend the definition proposed by Campbell and Hodos (1970) to behavior.

To what extent can behavior be traced back through a genealogical series to a stipulated common ancestral precursor? Would the result in any way be meaningful? Atz (1970) has discussed the concept of homology and its applicability to behavior. He assumes that homology is a concept that relates to structural correspondence of characters and is only meaningful in behavior to the extent that behavior can be related to or conceptualized as structure. Although behaviorists do use structural concepts, they are abstractions and not physically measurable in a direct way. Therefore, Atz reasons, the usefulness of the term "homology" in behavioral studies depends on the extent to which structural correlates of behavior can be found. However, because known structural correlates of behavior are few at present, he does not recommend the use of the term in behavioral research at this time. In contrast, Atz points out that the existence of homologous behaviors cannot be denied, for that would deny that behavior is a characteristic of organisms that is subjected to evolutionary change.

As Atz has already provided a scholarly and thoughtful survey of past and present applications of the homology concept to behavior, I shall not attempt to duplicate it here. Instead, I shall begin with his assumption that the recognition of homologous behaviors requires a structural correlate. This conclusion seems to be either implicit or explicit in virtually all discussions of the evolution of behavior. One of the earliest of these was Darwin's *The Expression of Emotions in Man and Animals* (1899). The following excerpt gives the flavor of Darwin's thoughts on this subject:

> With respect to the involuntary bristling of the hair, we have good reason to believe that in the case of animals this action, however it may have originated, serves, together with certain involuntary movements, to make them appear more terrible to their enemies; and as the same involuntary and voluntary actions are performed by animals

nearly related to man, we are led to believe that man has retained through inheritance a relic of them, now become useless. It is certainly a remarkable fact, that the minute unstriped muscles, by which the hairs thinly scattered over man's almost naked body are erected, should have been preserved to the present day; and that they should still contract under the same emotions, namely terror and rage, which cause the hairs to stand on end in the lower members of the Order to which man belongs [pp. 307–308].

BEHAVIORAL HOMOLOGY AND HOMOLOGY OF STRUCTURE

Behavior does not occur in the absence of structure. Behavior requires structural correlates and these structural correlates are the basis of an organism's behavioral heritage. Structural correlates of behavior consist of both such peripheral structures as nerves, muscles, glands, bones, etc., and such central nervous system structures as specific populations of neuron types, dendritic configurations, neurotransmitters, specific connections between neuronal populations, etc. Therefore, when we say that two behaviors are homologous, we are really saying that they depend on homologous structures; that is, they depend on structures that have a common phyletic origin. Lack of similarity of appearance or lack of similarity in function is not evidence against a homologous relationship between behaviors because such differences are quite compatible with the common ancestry of the structural substrates of the two behaviors. For example, the wings of birds and the forelimbs of mammals are homologous as derivatives of the forelimbs of the common reptilian ancestors of birds and mammals. To the extent that structural entities of common phyletic origin are used, any behaviors that involve use of the wings and use of the forelimbs are homologous, at least as movements. However, because of the considerable differences in structure and adaptive value between wings and limbs, behaviors involving these organs may have different appearances and may serve different functions.

A point that Ghiselin (1966) has stressed in discussing the concept of homology is that a statement that two structures are homologous that does not contain an explicit indication of the basis of the homology is essentially meaningless. For example, to say that a horse's front hoof is homologous with the middle finger of a primate's hand is an inappropriate description of the relationship between these two structures. A complete description of the phyletic homology must include the statement that these two structures are homologous as derivatives of the digits of the anterior appendages of early insectivores. Atz (1970) provides another illustration of this point: the anal fin of male four-eyed fishes (Anablepidae) is modified to form a gonopodium. The anal fin of the male of guppy and other members of the Poeciliidae is also modified to form a gonopodium, but of a different form than the comparable organ in Anablepidae. In view of the evidence of the independent origin of the gonopodium in Poeciliidae and Anablepidae, Atz considered the anal fins in these two families to be homologous as fins, but not as gonopodia. This specification of the basis of the homology is crucial for the concept of behavioral homology, if we accept the thesis that behaviors can be regarded as homologous only to the extent that their structural substrates can be regarded as homologous. The reason that this specification of the basis of the homology is essential is that a given behavior may not have the same substrate throughout phylogeny. Two behaviors of two different species, no matter how identical they appear, cannot be regarded as homologous in the sense that they represent a common ancestral heritage if they are each mediated by structures without a common ancestral heritage. For example, a considerable amount of evidence suggests that both the peripheral receptors, nerves, central cell groups and fiber pathways that mediate audition have undergone considerable change throughout vertebrate evolution (Ariëns-Kappers, Huber, & Crosby, 1960; Cahn, 1967; Tavolga, 1964; Wever, 1976, Chapter 5). The loss of the lateral-line contribution to acoustic reception, the progressive diminution of the lagena, the loss of associated nerves and central structures, as well as the progressive development of the cochlea and associated nerves and central structures suggest that hearing in mammals may have quite a different structural substrate from hearing in amphibians or bony fish. Similarly, on the motor side, sound production in marine mammals and bony fishes (Tavolga, 1964) involves completely different peripheral mechanisms than does sound production in terrestrial animals. Therefore, sound production in marine and terrestrial animals, even though of very similar acoustical properties and patterns, cannot be regarded as homologous if nonhomologous nerves and peripheral apparatus are responsible for generation of the sounds. Therefore, a definition of behavioral homology must include a statement of the underlying structural elements that at least in principle can be traced to specific ancestral precursors. I offer the following as a definition of behavioral homology:

> Behaviors are considered homologous to the extent that they can be related to specific structures that can, in principle, be traced back through a genealogical series to a stipulated ancestral precursor irrespective of morphological similarity.

Hunsaker (1962) has described the species-specific head-bobbing movement of various species of lizards of the genus *Sceloporus*. Although the movements differ from species to species in frequency, amplitude, duration, and pattern, they nevertheless are homologous because they all have a morphologically homologous substrate. In contrast, color change in African chameleons of the genus *Chamaeleo* is under the control of spinal nerves, whereas in American chameleons of the genus *Anolis* (actually small iguanids), color change is unaffected by sectioning of spinal nerves but is under the control of animal's endocrine system (Prosser & Brown, 1961). Therefore, very similar appearing behaviors, even in relatively closely related forms (in this case members of the same suborder) (Goin & Goin, 1971), may have a different structural basis not traceable to a common ancestral precursor and hence are not homologous as defined here.

LEVELS OF BEHAVIORAL HOMOLOGY

Many different behavior patterns have a considerable number of structures in common; e.g., the variety of behavior patterns that utilize the forelimbs or the hindlimbs. Do we wish to state that a behavior in one species that is associated with a particular structural entity is behaviorally homologous with any behavior in another species that is associated with a homologous structural entity? The answer to this question depends on the sense in which the term "homology" is used. For example, head scratching in birds, which is accomplished with the hindlimb, is homologous as a movement with head scratching in dogs, which is also accomplished with the hindlimb. Neither of these is homologous with head scratching in man, which is carried out with the forelimb. However, do we want to say that head scratching in dogs, which is carried out with the hindlimbs is homologous with running in birds, which is also carried out with the hindlimbs? At one level, the answer must be "yes" because of the homologous relationship between the hindlimbs of birds and mammals. This relationship is so general (that is, so many other behaviors also depend on these structures), however, that it is of little value in reconstructing the evolutionary history of either head scratching or running. Even in the example of head scratching in birds and dogs, the homologuous relationship of the behavior in the two species as movements does not answer the real question of the history of head scratching in the lines of descent leading to modern dogs and modern birds. In order to answer this question, it is necessary to know whether the Cotylosaurs scratch their heads with their hindlimbs. If evidence can be adduced from their living descendents and the fossil record that indicates that they did, then it can be stated that head scratching in birds and dogs is not only homologous as a movement but also homologous as head scratching.

SPECIAL CASES OF HOMOLOGY

What of those situations in which only a portion of a given behavior pattern of two species seems to have a homologous structural basis? Can homologous behaviors be spoken of in such instances? They can, but only with regard to that portion of the total behavioral pattern that has the homologous substrate. Such a case may be termed a "fractional homology." An example of a fractional homology may be seen in the feeding behavior of pit vipers in which the pit organ is used to detect the infrared radiation of prey (Goin & Goin, 1971) and the feeding behavior of ducks as they sift through the mud for prey (Wallace,

1963; Welty, 1963). Although the snakes use distance reception and the ducks use contact reception, the two behaviors are fractionally homologous. The common structural element is the trigeminal nerve, which can readily be traced to the trigeminal nerves of Cotylosaurs, the presumed common reptilian ancestors of birds and reptiles. In addition, since we have good reason to believe that the reptilian ancestors of snakes and birds have used their trigeminal nerves to mediate feeding, we can state that these aspects of the feeding patterns of ducks and pit vipers are also fractionally homologous as feeding behaviors.

Another special case of homology is "field homology." When a structure such as the spinal column, consists of many parts, one may not be able to trace any individual part to a common ancestral precursor, although the total aggregation of parts can be so traced. Therefore, although a specific vertebra of the spinal column of a snake cannot be related to a specific vertebra of the spinal column of a hippopotamus, there is little doubt that the spinal columns of these two vertebrates have been derived from the spinal columns of their common reptilian ancestor. The concept of field homology must be considered in any analysis of behaviors in terms of their basis in homologous structures. The studies of locomotion and intention movements by Daanje (1950) seem to be examples of behavioral field homology. The evidence seems plausible that the various intention movements are derived in general from the various locomotor movements; however, in many instances no specific element of the ritualized intention pattern can be related to a specific element of the locomotor pattern.

Atz (1970) raises the question of the extent to which the products of behavior, such as nests and vocalizations, can be regarded as representing homologous behaviors. He points out that the assumption that similar products are necessarily produced by homologous behaviors is unwarranted. However, if an analysis of the method of producing the product were to indicate that homologous structures were involved, then the products would indeed be homologous. The sound productions of marine and terrestrial mammals, as indicated above, would not be homologous, whereas the vocalizations of carnivores and primates would be homologous. Similarly, nest building that involved the use of homologous limbs would be homologous. However, nest building using nonhomologous limbs or involving oral manipulation of nest materials as compared to digital manipulation would not be homologous behaviors.

BEHAVIORAL HOMOPLASY

The example cited above of chromatophore regulation having a nonhomologous structural basis in certain lizards is a good illustration of behavioral homoplasy, i.e., similarities in behavior that are associated with nonhomologous structures. The following is suggested as a definition of behavioral homoplasy:

> Behaviors that are similar but that cannot be related to structures that can, in principle, be traced back through a genealogical series to a stipulated common ancestral precursor are homoplastic.

Several processes can contribute to behavioral similarity in the absence of homologous structural correlates: among these are convergence, parallelism, and mimicry.

Convergence is the process by which animals without a common ancestry pertinent to the characteristic in question develop similar structures in response to similar pressures from the environment. Hinde (1970) describes an interesting example of convergence in structure and behavior in two species of birds, the kittiwake (*Rissa tridactyla*) the gannet (*Sula bassana*). The kittiwake is a member of the order Charadriiformes, which includes the other gulls and the shore birds. The gannet is a member of the order Pelicaniformes, which includes the pelicans and cormorants. Both the gannet and the kittiwake nest on high cliffs, quite unlike other members of their respective orders. The kittiwake has developed a number of behavioral adaptations to this type of environment that differentiate it from its ground-nesting gull relatives. The gannet, which is not a particularly close relative of the kittiwake, resembles the kittiwake behaviorally in many of the very behavior patterns that differentiate the kittiwake from its closer relations among the gulls.

Both the distribution and variety of mechanisms of chromatophore control among the various vertebrate classes (Prosser & Brown, 1961) suggest that those animals that possess the ability to change color have acquired this ability independently and not as a heritage from remote ancestors. The presence of color change controlled by different mechanisms in closely related forms, such as different species of lizards, illustrates another process by which animals of common ancestry and encountering common problems of survival independently develop similar structures. In the case of the lizard chromatophores, the difference in control mechanism argues strongly for a homoplastic rather than a homologous relationship. However, often the similarity is so complete that only knowledge of the condition in the common ancestor or a study of the distribution of the character among extant forms can differentiate between similarity resulting from homology and similarity resulting from parallelism.

Another mechanism that results in homoplasy is mimicry. Wickler (1968) describes many examples of morphological mimicry in which an organism develops resemblance to another organism as a result of the adaptive value of the resemblance. Common examples are insects that resemble plants, nonpoisonous insects that resemble poisonous insects, etc. Although Wickler cites relatively few behavioral mimicries, the illustrations are instructive. For example, he describes the zone-tailed hawk (*Buteo albonotatus*), which glides in the midst of a flock of vultures. Because vultures are scavengers rather than predators, their presence does not evoke fear reactions from small animals as does the appearance of a hawk. The zone-tailed hawk, which also bears a morphological resemblance to the vultures, suddenly dives from the vulture flock and seizes the unsuspecting prey. Perhaps the most widely known example of behavioral mimicry can be found in the ability of birds, such as the mockingbird (*Mimus polyglottos*) and the brown thrasher (*Toxostoma rufum*), to imitate the calls of other species. Welty (1963) describes the snakelike hisses emitted by many hole-nesting species of birds and the monkeylike whistles emitted by the African crowned eagle (*Stephanoaëtos coronatus*) that attract the monkeys on which this bird preys.

A term that has been associated with homology and homoplasy is "analogy." Analogy refers to similarities in function. Because participation in behavior can be a function of a structure, is not behavior itself a special case of analogy? The answer to this question is "no." Although, in a sense, behavior may be said to be a function of a structure, the behavior is not an end in itself but instead serves a function to the animal. Therefore, analogous behaviors are defined here as "correspondence between behaviors of similar function, whether or not they can be related to homologous structures." The insertion of food into the mouth by a man and a monkey would be both homologous (because the hands of monkeys and humans are derived from the hands of their common ancestors) and analogous (because the behaviors serve the same purpose). The comparable behaviors in a racoon and a monkey are homoplastic with respect to the hands (because the hand structures of monkeys and racoons are the result of convergence rather than common ancestral heritage), homologous with respect to the derivation of the forelimb and appendages from common ancestors, and analogous because the behaviors serve the same function. However, the use of the trunk to insert food into the mouth by an elephant represents behavior that is analogous to food insertion with the forelimb digits by racoons, monkeys, and humans but is neither homologous nor homoplastic, because an elephant's trunk has a totally different derivation from that of the forelimb and digits and bears little if any resemblance to manual insertion of food.[1]

Behavioral homology has been defined here in the context of specific structural entities that are homologous. A point on which there seems to be general agreement is that behaviors that are associated with nonhomologous structures cannot be regarded as homologous, no matter how similar they appear (Atz, 1970; Tinbergen, 1962). However, Atz (1970) states that the converse is not necessarily true: "the fact that homologous structures are involved in behaviors may be an indication that the behaviors are homologous, but it cannot be considered proof [p. 60]." He bases this reservation on the contention that

[1] The food insertion behavior of elephants may be homoplastic with that of racoons and primates from one point of view. The rostral tip of an elephant's trunk (with some stretch of the imagination) bears a resemblance to a hand encased in a mitten and is used to grasp small objects in a similar manner.

homologous structures may not have homologous functions. There seems little doubt that the functions of particular structural entities change throughout the evolutionary history of a given lineage. However, the fact that two structures of common phyletic origin in two different species may have different functions in no way alters the derivation from a common precursor that is denoted by saying that they are homologous. Behaviors associated with the two homologous structures would therefore be homologous no matter how different the functions of the two behaviors were. A similar view was expressed by Tinbergen (1959).

APPLICATIONS OF THE CONCEPT OF BEHAVIORAL HOMOLOGY

Movements are a class of behavior closely related to structural entities. In the broadest sense, one can state that any movements produced by homologous structures are homologous. The movements of an anal fin and a gonopodium of fishes are therefore homologous because of the homologous nature of anal fins and gonopodia as anal fin derivatives, even though the movements of each are totally different in appearance and even though the two structures are not analogous (that is, differ in function). Similarly, Tinbergen (1959) has analyzed the various displays of gulls in terms of the specific structural correlates (head, bill, carpal joints, neck, etc.) and concludes that given the assumption that gulls are a monophyletic group, then the various displays observed in different species must be the result of divergent evolution from a common starting point.

Atz (1970) has pointed out that the most convincing examples of homologous behaviors are those that are uniquely observed in closely related species. On this point he and I agree completely. However, he goes on to state that " . . . as one goes beyond such a patent, ideal situation and begins to deal with less closely related species and with behavior patterns that are not unique and whose distribution is less coherent, the certainty with which the behavior can be homologized rapidly diminishes" (Atz, 1970, p. 66). Here he and I are in complete disagreement. The degree of phyletic relatedness is not the issue in behavioral homology as defined here. The issue is the degree to which behavior can be related to specific structural entities. Whales and birds are surely of rather remote common ancestry; yet the homology of the whale's flipper and the bird's wing as derivatives of the forelimbs of stem reptiles is well established. The certainty with which behavior can be homologized rapidly diminishes when we are unable to relate specific behavior patterns to specific structural entities. Fixed action patterns fall into the former category, in which specific behaviors associated with specific homologous structures are uniquely observed in closely related species. However, much of behavior, particularly mammalian behavior, is not fixed but plastic. Although many species-typical behaviors have been observed in mammals, such as aggressive and submissive displays (Dimond, 1970;

Johnson, 1972), courtship, and reproductive and parental behaviors (Noirot, 1972; Rowell, 1972), much of mammalian behavior, as well as the behavior of other vertebrates, involves the acquisition of new behavior, unique to the individual rather than to the species and not transmitted to successive generations; that is, changes in the behavioral phenotype without a change in the genotype. Such new behavior poses the greatest challenge to the evolutionary behaviorist because (1) such behaviors are often quite variable from individual to individual within a species; (2) no specific correlates can be found among peripheral structures (muscles, glands, limbs, integument, etc.) that differentiate one type of behavior from another; and (3) relatively little progress has been made in identifying the specific central structural correlates (specific neuronal populations and interconnections) of such behaviors. In a previous paper (Hodos, 1970), I have discussed the problems involved in comparing acquired behaviors across species and have suggested that a common reference point at which such behaviors can be compared is the maximum capacity to perform that behavior. All of the various difficulties involved in standardizing behavioral situations for animals with different degrees of development of various sensory and motor abilities, motivational level, seasonal effects, etc., can only serve to produce an underestimate of an animal's maximum capacity. Nothing short of a highly contrived circumstance, such as pharmacological intervention, could result in an overestimate of an animal's maximum. Therefore, if subsequent investigations, under more appropriate conditions, should yield a higher estimate of the maximum ability, the earlier estimates must be discarded. For example, I have suggested that the inferior performance of some animals, such as rodents, on visual learning-set tasks may be because vision is not these animals' best sensory modality and their performance on olfactory learning sets may be far superior. Recently, Slotnick (1972; Slotnick & Katz, 1974) has observed that rats perform as well in olfactory learning sets as great apes and humans perform in visual learning sets. My suggestion here is that the search for maximum abilities may provide a solution to the problem of the individuality of acquired behaviors in the reconstruction of behavioral history. The classic studies by Tryon (1929) in the genetics of maze learning indicate that although specific information is not transmitted genetically, the potential for efficient or inefficient acquisition, storage, retrieval, and utilization of information can become incorporated into the genotype in the same manner as body size, neck length, or any other hereditary trait.

The problem of specific peripheral structures associated with specific acquired behaviors must be circumvented altogether, because so many acquired behaviors make use of the same peripheral structures. The study of the evolution of the capacities to acquire new behaviors leads inevitably into the central nervous system as the locus of homologous structural correlates through which homologous behavioral capacities can be identified. Similar views have been expressed by Mayr (1958), Pribram (1958), and Simpson (1958). Atz (1970) is also in basic agreement with this view but raises several points of caution: (1) relatively little is known about the specific neural correlates of learning, memory, perception, sensory–motor processes, etc.; (2) "how can one ever hope to confidently assume that any particular behavior has continuously been associated with a particular condition (whatever it may be) of the central nervous system?" [p. 62]. The first point is indisputable, as a general statement, but is less correct with regard to sensory systems, in which considerable progress has been made in identifying comparable cell groups and fiber pathways in a wide variety of representatives of all the vertebrate classes. Moreover, we seem to be on the threshold of an era of more rapid progress in uncovering the roles of various brain structures in these behaviors. However, the issue of the function of a structure or its role in behavior is not directly relevant to its homologous relationship (or lack of it) to another structure. All that is relevant is the derivation (or lack of it) of the two structures from a common ancestral precursor. The same principle applies to brain structures and behavior. Behaviors associated with brain structures that have a common genealogical history are homologous, whether or not the behaviors are of the same type or serve the same function to the animal. For example, the maleus and incus of the middle ears of mammals are derived from the quadrate and articular bones of the jaws of ancestral reptiles (Romer, 1966). The stapedius and tensor tympani muscles that control the articulation of these ossicles during the acoustic reflex are innervated by branches of the trigeminal and facial cranial nerves (Gulick, 1971), which is in keeping with their former role in feeding reflexes. On the one hand, to accept the homologous relationship of the maleus and incus of living mammals to the quadrate and articular bones of living reptiles and, on the other hand, not to accept the homologous relationship of mammalian acoustic reflexes to reptilian feeding reflexes would seem to be inconsistent. The fact that the two reflexes are different and serve different functions for the animal does not alter the fact of their common origin in ancestral forms. With regard to Atz's second point of caution, we do not have to assume that a given behavior has been continously associated with a particular brain structure throughout its evolutionary history. We merely have to assume that some behavior has been associated continuously with the structure. Whether the behavior has been modified by changes in adaptive requirements does not alter its ancestry.

BEHAVIOR AND SYSTEMATICS

A factor that seems to have injected considerable confusion into discussions of the evolution of behavior and behavioral homology is the use of behavior as a taxonomic character. Avian taxonomy has made particular use of behavioral characters in the classification and differentiation of species (Mayr, 1958). For example, the mockingbirds and thrashers are classified in the family Mimidae (Welty, 1963), which is differentiated from related families in part because of the behavioral characteristic of mimicry of the calls of other birds. Others have sought to use behavior as a means of determining evolutionary relationships (Lorenz, 1941). However, the aims of a taxonomist may not be the same as those of a phylogenist, except in the case of a purely phylogenetic taxonomy (Crowson, 1970). Classification according to similarity is a meaningful endeavor but may not result in a classification that is related to evolutionary descent. Caution must be employed to avoid the circular logic of using similarity of behavior to establish taxonomic categories and then using similarity of behavior within the established categories as evidence of behavioral homology. Some accounts of the evolution of display movements (Hinde & Tinbergen, 1958; Hinde, 1970) may contain elements of this circularity. Hinde (1970) has recognized the potential for circularity and urges the widest possible range of independent sources of evidence of phylogenetic relationships as a means of avoiding this logical trap.

BEHAVIORAL HOMOLOGY AND MOSAIC EVOLUTION

A concept that has important bearing on evolutionary behaviorism is the concept of "mosaic evolution" (Olson, 1971). "Mosaic evolution" refers to the fact that evolution does not progress at the same rate in all systems of an organism. Every organism is a mixture of both ancestral (slowly evolving) characters and derived (more rapidly evolving) characters. Therefore, there are no "primitive" animals. There are, however, animals in which ancestral characters predominate and these are the most useful as representatives of ancestral forms. However, one must not assume, a priori, that all behaviors found in such animals are necessarily representative of the behavior of the ancestral form. Before one can conclude that the observed behavior represents an ancestral condition, the structural substrate of that behavior must be demonstrated to exist essentially in the ancestral condition, rather than in a derived condition.

A related point is that the terms "ancestral" and "derived" are relative. Every ancestral condition was at one time derived from a still more ancestral condition. An ancestral condition is therefore not necessarily a rudimentary condition. Many extinct forms achieved high degrees of specialization of certain systems. Therefore, evolutionary trends should not be expected necessarily to occur as increasing monotonic functions.

LIMITATIONS OF THE CONCEPT OF BEHAVIORAL HOMOLOGY

Atz (1970) has summarized his views on the relationship between behavioral homology and homologous structures as follows: "when one thinks of homologous behavior, one is perforce thinking of behavior in morphological terms (fixed action patterns and the like) or as closely associated with some structure (neural or otherwise) [p. 68]." He concludes that the concept of behavioral homology is at present of little or even negative value because of difficulties and oversimplifications that arise from treating behavior as structure and because we seem to be unable to relate many behaviors to specific neural structures. Although Atz and I are in essential agreement on both points, I do not agree that the concept of behavioral homology is of little use. Because as many behaviors cannot be related to particular central nervous system structures as may be desired at present is not sufficient reason to suspend the use of a concept that is essential for the reconstruction of behavioral history.

The concept of behavioral homology is totally dependent on the concept of structural homology. This state of complete dependence imposes a number of limitations on the behavioral homology concept. First, behavioral homologies stand or fall with their structural correlates. If evidence comes to light indicating that structures previously thought to be homologous are not homologous, then any behaviors associated with these structures must also be redesignated non-homologous. Second, the structural concept of homology is itself not free from controversy and criticism. As Campbell and Hodos (1970) have pointed out, both the phylogenetic approach to homology and the structural similarity approach to homology ask meaningful scientific questions. They are different questions, however, and can result in different answers. Other criticisms have been leveled at the concept of structural homology and the methods by which these homologies are determined (Nelson, 1970; Schaeffer, Hecht, & Eldredge, 1972). Finally, the identification of homologous behaviors requires the identification of specific homologous structures. For many behaviors, particularly acquired behaviors, few specific structural correlates have been found at present. The quest for such correlates must lead into the central nervous system. This means that, to a large extent, progress in reconstructing the evolutionary history of such behaviors must be governed by the rate of progress in neuroanatomy, neurophysiology, neurochemistry, neuropsychology, and neuroethology. Unless behaviorists are prepared to carry out their own comparative research studies in these fields in order to provide structural correlates, they must await the fruits of the labors of others before they can proceed.

In spite of these limitations, which are by no means minor, the concept of behavioral homology should not be abandoned or shelved, but should be used wherever possible. For, without this concept we will be unable to reconstruct behavioral history using the behavior of living animals. By recognizing the limitations of the concept, by carefully defining terms, and by specifying the structural basis of the behavior, however, the goal of a meaningful statement of behavioral evolution may be achieved.

ACKNOWLEDGMENTS

A preliminary draft of this chapter was read by James W. Atz, Michael T. Ghiselin, Jack P. Hailman, Wolfgang M, Schleidt, and Matthew Yarczower. I am grateful for their valuable comments and suggestions. The preparation of this chapter was supported by Grant No. EY-00735 from the National Eye Institute.

REFERENCES

Ariëns-Kappers, C. U., Huber, G. C., & Crosby, E. C. *The comparative anatomy of the nervous systems of vertebrates, including man.* New York: Hafner, 1960 (reprint of 1936 edition).

Atz, J. W. The application of the idea of homology to behavior. In L. R. Aronson, E. Tobach, D. S. Lehrman, & J. S. Rosenblatt (Eds.), *Development and evolution of behavior.* San Francisco: Freeman, 1970.

Cahn, P. H. (Ed.) *Lateral line detectors.* Bloomington: Indiana University Press, 1967.

Campbell, C. B. G. Morphological homology and the nervous system. This conference.

Campbell, C. B. G., & Hodos, W. The concept of homology and the evolution of the nervous system. *Brain, Behavior, & Evolution,* 1970, **3,** 353–367.

Crowson, R. A. *Classification and biology.* New York: Atherton, 1970.

Daanje, A. On locomotory movements in birds and the intention movements derived from them. *Behaviour,* 1950, **3,** 48–98.

Darwin, C. *The expression of the emotions in man and animals.* New York: Appleton, 1899.

Dimond, S. J. *The social behavior of animals.* New York: Harper and Row, 1970.

Ghiselin, M. T. An application of the theory of definitions to systematic principles. *Systematic Zoology,* 1966, **15,** 127–130.

Goin, C. J., & Goin, O. B. *Introduction to herpetology.* San Francisco: Freeman, 1971.

Gulick, W. L. *Hearing: Physiology and psychophysics.* New York: Oxford University Press, 1971.

Hinde, R. A. *Animal behaviour.* New York: McGraw-Hill, 1970.

Hinde, R. A., & Tinbergen, N. The comparative study of species-specific behavior. In A. Roe & G. G. Simpson (Eds.), *Behavior and evolution.* New Haven, Connecticut: Yale University Press, 1958.

Hodos, W. Evolutionary interpretation of neural and behavioral studies of living vertebrates. In F. O. Schmitt (Ed.), *The neurosciences: Second study program.* New York: Rockefeller University Press, 1970.

Hodos, W. Comparative study of brain–behavior relationships. In I. J. Goodman & M. W. Schein (Eds.), *Birds: Brain and behavior.* New York: Academic Press, 1974. Pp. 15–25.

Hodos, W., & Campbell, C. B. G. *Scala naturae:* Why there is no theory in comparative psychology. *Psychological Review,* 1969, **76,** 337–350.

Hunsaker, D. Ethological isolating mechanisms in the *Sceloporus torquatus* group of lizards. *Evolution,* 1962, **16,** 62–74.

Johnson, R. N. *Aggression in man and animals.* Philadelphia: W. B. Saunders, 1972.

Källén, B. Embryological aspects of the concept of homology. *Arkiv för Zoologi,* 1959, **12,** 137–142.

Lorenz, K. Vergleichende Bewegungsstudien an Anatinen. *Journal für Ornithologie,* 1941. **89,** 194–294. Translated by C. H. D. Clarke with new revisions by Lorenz in P. H. Klopfer & J. P. Hailman (Eds.), *Function and evolution of behavior: An historical sample from the pens of ethologists.* Reading, Mass.: Addison-Wesley, 1972.

Mayr, E. Behavior and systematics. In A. Roe & G. G. Simpson (Eds.), *Behavior and evolution*. New Haven: Yale University Press, 1958.

Mayr, E. *Populations, species and evolution.* Cambridge, Mass.: Harvard University Press, 1970.

Nelson, G. J. Outline of a theory of comparative biology. *Systematic Zoology*, 1970, **19**, 373–384.

Noirot, E. The onset of maternal behavior in rats, hamsters and mice: A selective review. In D. S. Lehrman, R. A. Hinde, & E. Shaw (Eds.), *Advances in the study of behavior.* New York: Academic Press, 1972.

Northcutt, R. G. Discussion of the preceding paper. In J. M. Petras & C. R. Noback (Eds.), *Comparative and Evolutionary aspects of the vertebrate central nervous system. Annals of the New York Academy of Sciences,* 1969, **167**, 180–185.

Olson, E. C. *Vertebrate paleozoology.* New York: Wiley-Interscience, 1971.

Pribram, K. Comparative neurology and the evolution of behavior. In A. Roe & G. G. Simpson (Eds.), *Behavior and evolution.* New Haven: Yale University Press, 1958.

Prosser, C. L., & Brown, F. A. *Comparative animal physiology.* Philadelphia: W. B. Saunders, 1961.

Romer, A. S. *Vertebrate paleontology.* Chicago: University of Chicago Press, 1966.

Rowell, T. E. Female reproduction cycles and social behavior in primates. In D. S. Lehrman, R. A. Hinde, & E. Shaw (Eds.), *Advances in the study of behavior.* New York: Academic Press, 1972.

Schaeffer, B., Hecht, M., & Eldredge, N. Paleontology and phylogeny. In T. Dobzhansky, M. K. Hecht, & W. C. Steere (Eds.), *Evolutionary biology,* Vol. 6. New York: Appleton-Century-Crofts, 1972.

Simpson, G. G. Behavior and evolution. In A. Roe & G. G. Simpson (Eds.), *Behavior and evolution.* New Haven: Yale University Press, 1958.

Slotnick, B. M. Olfactory reversal learning sets in rats. *Conditional Reflex,* 1972, **7**, 179.

Slotnick, B. M., & Katz, H. M. Olfactory learning-set formation in rats. *Science,* 1974, **185**, 796–798.

Tavolga, W. N. (Ed.) *Marine bio-acoustics.* Oxford: Pergamon Press, 1964.

Tinbergen, N. Comparative studies of the behaviour of gulls (Laridae): a progress report. *Behaviour,* 1959, **15**, 1–70.

Tinbergen, N. The evolution of animal communication–a critical examination of methods. *Symposia of the Zoological Society of London,* 1962, **8**, 1–6.

Tryon, R. C. Genetics of learning ability in rats. *University of California Publications in Psychology,* 1929, **4**, 71–89.

Wallace, G. J. *An introduction to ornithology.* New York: Macmillan, 1963.

Welty, J. C. *The life of birds.* New York: Knopf, 1963.

Wever, E. G. Origin and evolution of the ear of vertebrates. In R. B. Masterton (Ed.), *Evolution of the nervous system and behavior.* Hillsdale, New Jersey: Lawrence Erlbaum Assoc., 1976.

Wickler, W. *Mimicry in plants and animals.* New York: McGraw-Hill, 1968.

12

Practical Aspects of Homology Recognition

Bobb Schaeffer

American Museum of Natural History

When turning from consideration of the conceptual and philosophical aspects of homology (Ghiselin, this volume) to the equally important problem of postulating specific hypotheses of homology, various practical difficulties may arise that cannot be resolved simply by the addition of a conditioning phrase (Bock, 1969). It may be stated that the shark mandible and the human mandible are homologous as vertebrate jaws, which is a correct statement, but this is not very meaningful in terms of the actual jaw components. It is usually easier to postulate homology in closely related animals than it is in more distantly related ones. The distinction between phylogenetic homology and topographic, or structural, homology seems to me unreal, as correspondence in relative position must often be used to identify the very structures that are being traced historically.

Although homology is usually considered in relation to the phylogenetic analysis of adult structures, it is important to recall that we can also deal with this phenomenon in relation to morphogenesis. Ectodermal, mesodermal, and ectomesenchymal cartilage are obviously homologous as cartilage regardless of their morphogenesis or where they are situated. Nevertheless, histologists have experienced difficulty in identifying and homologizing certain calcified tissues in fishes and tetrapods. A good example is the ectomesenchymal enameloid of most fishes, which can be distinguished from the ectodermal enamel on the teeth of tetrapods, mainly on the basis of development (Figure 1; see Moss, 1964; Poole, 1971). Is it meaningful to say that the enameloid of fish teeth and the enamel of tetrapod teeth are homologous as the most peripheral acellular layer of the dentition, or is greater exactitude desirable?

Most Tetrapods	Fishes and some Amphibians
Dentin induces ameloblasts to form enamel	Odontoblasts produce dentine below enameloid
Odontoblasts form dentine	Combined activity of IDE cells and odontoblasts responsible for enameloid
IDE cells (ameloblasts) induce subadjacent cells of dental papilla to change into odontoblasts	IDE cells may induce subadjacent cells of dental papilla to change into odontoblasts
Inner epithelium of enamel organ differentiates into ameloblasts	Cells of inner dental epithelium elongate
Dental papilla induces overlying ecoderm to form enamel organ	Dental papilla presumably induces overlying ectoderm to form dental follicle

Oral epithelium induces neural crest cells to form dental papilla

Neural crest cells become associated with oral epithelium

FIG. 1 Hypothetical reconstruction of morphogenetic pathways for dental tissues in fishes and tetrapods.

There are numerous well-known examples of homology at the adult organ level, such as the endostyle and thyroid gland, or certain bones of the reptilian upper and lower jaws and the middle ear ossicles of mammals. These are clear cut examples in spite of change in function. The problem of homology becomes particularly sticky when structures in, for instance, the urochordates and the cephalochordates are compared with those in the vertebrates. The branchial basket of a tunicate and that of *Amphioxus* arise from mesoderm. In all vertebrates the branchial skeleton develops from neural crest cells, or ectomesenchyme. The nonvertebrate branchial apparatus is composed of cartilage-like tissue, whereas that of the vertebrates always at least begins ontogenetically as true cartilage. Although it may be said that all of these branchial skeletons, with their so-called "gill slits," are homologous as skeletal supports of the pharyngeal cavity, there is a fundamental distinction in morphogenesis. Perhaps this example can be regarded as one in which homologous adult structures have evolved along partly different developmental pathways.

Another rather similar homology question involves the development of ossified centra in the vertebral column of certain related groups of higher bony fishes. There is fairly good paleontological evidence that the earliest adult representatives of three of four families of halecomorph fishes (Patterson, 1973) have had persistent notochords that were not surrounded or replaced by ossified tissue in the adult stage (the primitive character state). Nevertheless, more recent members of the same families independently developed perichordially ossified centra that partly or completely replaced the notochord (Schaeffer, 1967). This is perhaps what de Beer (1951) has called latent homology—a condition in which the manifestation of a character in a common ancestor is obscured or delayed for developmental reasons. Presumably, the sclerotomic tissue surrounding the notochord in the common ancestor of these families and in the early representatives of each family possessed the potentiality for producing bone. For some reason this potentiality did not become an actuality until some unknown morphogenetic mechanism was activated.

The point I wish to make is that homology decisions are not always meaningful if only adult stages can be compared. Without wishing to overemphasize this point, I think that many homology problems cannot be dealt with effectively until more is known about the morphogenetic factors involved. For example, the snout—dermal bone pattern in each of three closely related taxa of fossil ray-finned fishes is distinctive (Figure 2). On the basis of the available samples there is no demonstrable variability within each taxon, and there is no tendency for one pattern to approach another (Schaeffer, 1967). The snout pattern in which the postrostral bone completely separates the nasal bones on each side is considered the primitive condition. The pattern in which the enlarged nasals meet in front of a reduced postrostral is structurally intermediate. A third arrangement, in which the nasals meet in the midline and in which the postrostral is totally absent, may be regarded as the most derived, or specialized, of the three. The reduction and loss of the postrostral may have involved competition between the postrostral and the nasal ossification centers, and the mesenchyme that usually forms the postrostral may have been taken over entirely by the nasal ossification centers. Examples of apparent bone multiplication, fusion, or elimination are quite common in early groups of higher bony fishes. There has been a tendency on the part of some workers to use compound bone names when fusion is thought to have occurred, or to call a number of small bones by the same name as a single large one when the small bones occupy essentially the same area in a different taxon. In point of fact, there is no information about how ossification centers interact in utilizing available mesenchyme for dermal bone ossification and, therefore, no realistic or meaningful approach at present to the homology of such bones.

A final example is related to the origin of the tetrapod pectoral and pelvic limbs. Although the humerus and femur, and perhaps the radius, ulna, tibia, and fibula, can be homologized with the proximal elements of the rhipidistian

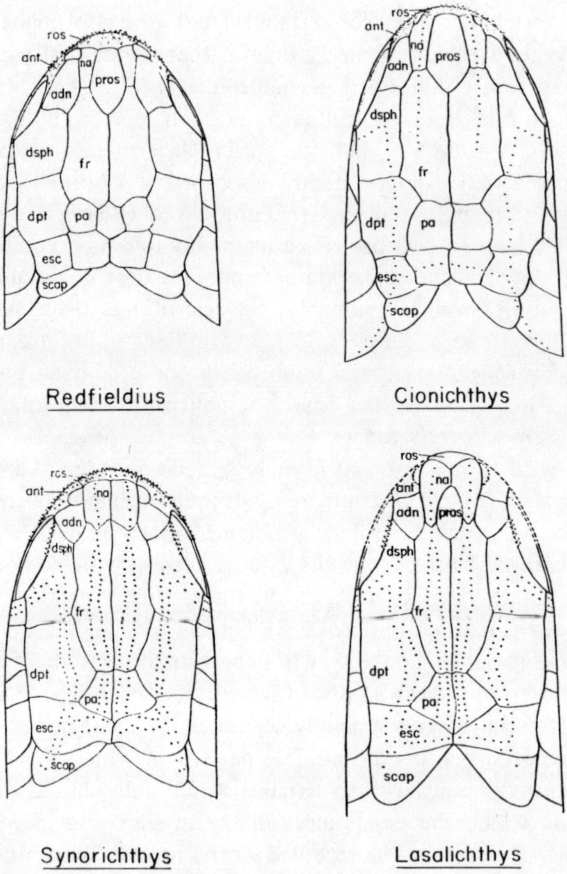

FIG. 2 Skull roofs in four genera of Triassic redfieldiid fishes. Abbreviations: adn, adnasal; ant, antorbital; dpt, dermopterotic; dsph, dermosphenotic; esc, extrascapular; fr, frontal; na, nasal; pa, parietal; pros, postrostral; ros, rostral; scap, suprascapular.

crossopterygian pectoral and pelvic fin skeletons, the resemblances disapper more distally (Andrews & Westoll, 1970). There must have been a major morphogenetic change to produce the carpus, tarsus, and phalanges. At best, the homologies for the more distal parts of the fins and limbs are so general as to be almost meaningless.

In regard to recognizing homologous subsystems, pathways, nuclei, and parts of the central nervous system, I can only reiterate something I said at the 1969 conference on the vertebrate central nervous system (Schaeffer, 1969): "Comparative study of brain archticture in fishes [and for that matter, in many other groups of living vertebrates] frequently includes taxa that are separated from their common ancestry by tens to hundreds of million years." For this reason the recognition of homologous structures within the brain may be exceedingly difficult.

In conclusion, most investigators (whether they admit it or not) assume that two or more structures are homologous unless there is reason to believe otherwise. This operational approach promotes comparison and generates hypotheses that might be otherwise neglected. Nevertheless, I believe that testable experimental evidence can be particularly helpful in providing new insights into a variety of homology problems, as in the comparison of enameloid and enamel. It seems obvious that the branchial skeleton of *Amphioxus* is homologous with that of vertebrates but to be meaningful a hypothesis about this homology should include a statement about the difference in morphogenesis.

Although the philosophical aspects of homology are being reworked, consideration of the morphogenetic aspect has not kept pace. For assessing the homology of many characters used by systematists, topographic relationship is the main criterion available. The very nature of homology implies a level of understanding that rarely exists but which, nevertheless, is clearly desirable. The ways in which enamel and enameloid are homologous can now be stated with some precision because there is a testable hypothesis regarding the morphogenetic pathway for each.

REFERENCES

Andrews, A. M., & Westoll, T. S. The postcranial skeleton of *Eusthenopteron foordi* Whiteaves. *Transactions of the Royal Society of Edinburgh,* 1970, **68**(9), 207–329.

Bock, W. J. The concept of homology. *Annals of the New York Academy of Sciences,* 1969, **167,** 71–73.

de Beer, G. R. Embryos and ancestors. Oxford: Clarendon Press, 1951.

Moss, M. L. the phylogeny of mineralized tissues. In W. J. L. Felts & R. J. Harrison (Eds.), *International review of general and experimental zoology,* Vol. 1. New York and London: Academic Press, 1964. Pp. 297–331.

Patterson, C. Interrelationships of holosteans. In P. H. Greenwood, R. S. Miles, & C. Patterson (Eds.), Interrelationships of fishes. *Journal of the Linnean Society (Zoology),* 1973, suppl. no. 1, **53**, 233–305.

Poole, D. F. G. An introduction to the phylogeny of calcified tissues. In A. Dahlberg (Ed.), *Dental morphology and evolution.* Chicago and London: University of Chicago Press, 1971, pp. 65–79.

Schaeffer, B. Osteichthyan vertebrae. *Journal of the Linnean Society (Zoology),* 1967, **47**(311), 185–195.

Schaeffer, B. Adaptive radiation of the fishes and the fishnamphibian transition. *Annals of the New York Academy of Sciences,* 1969, **167,** 5–17.

13

In Defense of the Analog: A Commentary to N. Hotton

Stephen Jay Gould

Harvard University

The Manichean classification of organic resemblances divides all similarities of form into treacherous, deceptive analogs and guiding, enlightening homologs. As in all such distinctions, the good and the evil acquire their status only with regard to specified goals and beliefs. In this case, the basis for classification is a dogma that proclaims phylogeny as the supreme goal of evolutionary biology—for homologous features are true marks of common descent, whereas analogs only masquerade as signs of ancestry. Thus, many great advances of nineteenth century comparative anatomy and embryology involved the identification as analogous of structures previously judged homologous. The vertebrate skeleton, for example, is not homologous with the arthropod carapace; and insects do not, as St. Hillaire had argued, dwell within their own vertebrae.

This prestige of the homolog has dictated an approach to a dilemma that paleontologists continually face: fossils cannot perform and behave. In moving beyond tenuous inferences from morphology and sedimentary environment, we must study the living organism "most like" the designated fossil. Almost automatically, "most like" is defined by homology as the living organism sharing most recent common ancestry with the designated fossil. This is done despite vast differences in life styles and immense separations in time (lampreys as models of cephalaspids, for example). Dr. Hotton, however, has advanced the faintly heretical notion that it may be better to use the closest analog for the comparisons (tadpoles for cephalaspids, for example). I should like to defend this suggestion and to extend its scope by presenting three general arguments in favor of the much abused analog. The establishment of phylogeny does not, after all, exhaust the content of evolutionary biology.

THE ANALOG AS "MOST LIKE"

I am inclined to agree with the numerical taxonomists and to argue that convergence generally affects only a few characters; therefore, for the totality of morphology, physiology, and behavior, the animal "most like" a fossil is probably its closest living neighbor on a phyletic tree. However, when such particular organs as the brain are studied—and especially when general function rather than detailed morphology is considered—I suspect that an analog can provide more enlightening comparisons. By "analog" I mean an organ or organism of the same *Bauplan* that shares with the fossil a similarity in what it does rather than in where it has come from. If a feature is functionally conditioned by habitus rather than passively retained via heritage, then study the analog rather than the homolog. The constraints of optimal design are sufficiently stringent that similar habitus usually requires a similar adaptive response through parallel or convergent evolution. In this context, the amphibian tadpole may well provide more insight into the function of the cephalaspid brain than *Petromyzon* attached to its host in the Great Lakes.

To choose an example from my own field: if an understanding of the functional morphology of swimming in squids and scallops is desired, another mollusc of standard design would be among the worst standards for comparison. Squids are functionally like vertebrates and should be so treated (Packard, 1972). In interpreting the hydrodynamic adaptations of scallops, the proper functional analog is not even a standard aquatic teleost (much less a benthic clam) but rather a bird, an airplane, or another machine that must generate lift in a medium lighter than itself (Gould, 1971a).

A related problem arises when homologies are truly sought but it is then uncritically assumed that they can be provided by closest phyletic relatives. Bauchot and Stephan (1966, 1969), for example, have based a long series of papers on the assumption that a brain–body curve for "basal insectivores" represents a "primitive" allometric trend for mammals in general. This curve is based primarily on a contrast in the relative brain size of small shrews versus large hedgehogs. Now, if gross brain size—and this means simple mass, not the subtleties of detailed morphology—is more a response to immediate conditions of life in these modern insectivores than a remembrance of things past, then the brain–body curve of "basal insectivores" is unlike that of mammalian progenitors, even though these insectivores are the closest phyletic relatives of ancestral mammalian groups. The issue is of some importance because Bauchot and Stephan base their assessments of brain size on ratios between brain weights of modern mammals and expected values for "basal insectivores" of the same body weight. Because the expected basal insectivore value is often a distant extrapolation from the modest size of a large hedgehog, the problem becomes acute if the calculated basal insectivore slope of 0.63 departs even slightly from true ancestral values. (If the true value were even so close to 0.63 as the commonly accepted 0.67, significant differences would result from extrapolation to the size of, say, a cow.)

THE ANALOG AS A REPLICATE

The techniques of evolutionary biology are often said to be descriptive, holistic, and historical, in contrast with the methods of experiment and analysis that have insured the spectacular success of other fields. This contrast is partly true, and that part should be defended and even cherished; nonetheless, experiment is so powerful a tool that we are well advised to use it when possible.

The analog is the paleontologist's only source of experimental material; for it is a replicate (however imperfect) in the testing of any functional hypothesis. Homologous organs cannot be used in experiment, for similarity of function only reflects the unenlightening fact that the same organ is present in two or more creatures. However, the convergent evolution of similar structures fulfills, at least imperfectly, the criterion of independent replication that any experiment requires. An adequate theory of functional morphology must explain the constraints on adaptive design by studying how different organisms react to the same selective regime. If we want to know whether plate tectonics is a true, universal physics of large bodies or only a descriptive account of this planet's history, other planets must be studied. If we want to know whether the biochemical unities of all life on earth have general import as optimal designs (given the nature of universal chemistry) or merely reflect the monophyletic origin of life on earth (and the homologous status of ATP and left-handed amino acids), then we shall have to hope for life on Mars. If, to retreat to something more immediate, one wishes to assess the functional limits or mechanical constraints upon the human eye, one would do well—as J. Z. Young and others have done—to study the octopus.

THE ANALOG AS REINFORCER OF THE FUNCTIONAL THEME

No one doubts that Gregory's (1936) distinction of habitus and heritage represents the two determinants of organic form—an animal has a structure either because its ancestor has had it or because it has evolved it in response to requirements of its immediate habitat. Nonetheless, the phyletic dogma reigns to exalt heritage and relegate habitus; this bias is most strongly expressed by modern Hennigians, who treat analogy as Haeckels' *Fälschungsgeschichte,* something to be recognized and removed lest phylogeny be confounded. When Wilhelm His (1888, p. 295) did battle with Haeckel's extreme phyleticism, he criticized Haeckel's trees as "rigid morphological diagrams, abstracted by merely logical operations." He urged the infusion of functional thinking as an antidote to "pure" morphology and guarantor of a richer, more comprehensive biology. I do not object to phyleticism and the search for homology in its proper place–in the establishment of phylogeny; I do deplore the subtle extension, often implied, that analogous resemblance be generally categorized as treacherous–for analogy asserts the functional theme as well as homology reflects the phyletic.

Hotton has contrasted two modes of origin for vertebrate classes. He argues that the lower vertebrate classes arose by rapid invasion of new environments by forms that had expanded to an edge of the adaptive zone of ancestral types. New classes may then arise from single clades in virtual absence of competition with parental forms. This is the classic, monophyletic picture for the origin of higher taxa–shared (homologous) features result from shared descent from a single ancestral form.

The mammals, in contrast, arose by broad structural modification in several lineages within a habitat that was basically occupied, and therefore in the face of continuous competition. Putting aside the issue of how many lineages actually crossed a (largely arbitrary) threshold, I note that the evolution of mammals is replete with active, functionally conditioned, independent acquisitions of similar features. In this light, I must focus on the functional theme and view common features more as marks of similar selective regimes than as examples of simple retention from common ancestors. I find this perspective intriguing: what can be said about homologous resemblance beyond using it to construct phylogeny? With analogous resemblance, the nature of functional response can be studied as a natural experiment with replication. We can assess optimal design and the extent of its realization. We can understand why morphology does not range completely across the spectrum of conceivable form (Raup, 1966).

The functional theme has its extreme statement–as uncompromising as pure phyleticism and as far from Aristotle's *aurea mediocritas.* This is D'Arcy Thompson's vision of form produced directly by physical forces acting on "plastic" organic matter (Thompson, 1942; Gould, 1971b). To D'Arcy Thompson, external form reflects little more than the nature of surrounding forces and the intrinsic size of a creature (surface forces mold a small animal, gravitational forces a large one). Homology, and phylogeny itself, is a chimera because a graded series of form must reflect a gradual alteration of external forces, not a slow, Darwinian transmutation. This sounds so deliciously heretical; and yet, is it any further from a really adequate biology than the equally uncompromising phyleticism that still has the respectability of football and apple pie in many evolutionary circles?

REFERENCES

Bauchot, R., & Stephan, H. Données nouvelles sur l'encéphalisation des insectivores et des prosimiens. *Mammalia,* 1966, **30**(1), 160–196.

Bauchot, R., & Stephan, H. Encéphalisation et niveau évolutif chez les Simiens. *Mammalia,* 1969, **33**(2), 225–275.

Gould, S. J. Muscular mechanics and the ontogeny of swimming in scallops. *Palaeontology,* 1971a, **14**(1), 61–94.

Gould, S. J. D'Arcy Thompson and the science of form. *New Literary History,* 1971b, **II**(2), 229–258.

Gregory, W. K. Habitus factors in the skeleton of fossil and recent mammals. *Proceedings of the American Philosophical Society,* 1936, **76,** 429–444.

His, W. On the principles of animal morphology. *Proceedings of the Royal Society of Edinburgh,* 1888, **15,** 287–298.

Packard, A. Cephalopods and fish: the limits of convergence. *Biological Reviews*, 1972, **47,** 241–307.

Raup, D. M. Geometric analysis of shell coiling: general problems. *Journal of Paleontology,* 1966, **40,** 1178–1190.

Thompson, D. W. *On growth and form*. Cambridge University Press, 1942.

14

Homology: Logic, Information, and Efficiency

Jack P. Hailman

University of Wisconsin

Because "homology" is a scientific term and therefore a vehicle of communication, no definition of it is right or wrong, but thinking makes it so. The history of the term (e.g., Ghiselin, 1969, and Chapter 10 of this volume) or its use by various authors (e.g., Campbell & Hodos, 1970) can be sought, but neither sort of investigation yields a definition that must be followed. Unlike taxonomic names, scientific terms have no rules of priority, and I think it undesirable to let the past become the tyranny of present thinking. However, the logical content of scientific concepts can be examined and their efficiency as carriers of information determined–the task I undertake here for such concepts as "homology," "analogy," "homoplasy," and their kin.

PRE-DARWINIAN ORIGINS

As reviewed by Ghiselin (1966, 1969, and Chapter 9 of this volume) the concepts of homology and analogy derive from Richard Owen's (1843) use of "homolog" and "analog." It is tempting but inappropriate to read evolutionary meaning into pre-Darwinian conceptions. Owen, like most of his contemporaries, appears to have believed that God created animals on a limited number of blueprints, which are basically Platonic essences or ideals. Owen apparently felt that by careful investigation of the characteristics of animals the basic blueprint could be discovered through comparisons. Therefore, similar characters, which are merely construction variants of the blueprint, are "homologs." Characters that may be mistaken for "homologs" but that have not been built from the same blueprint, are "analogs." As Ghiselin (Chapter 9 of this volume) shows, Owen is inconsistent in his use of "analog," defining the resemblances, on the one hand, as resulting from the common function the characters must serve and on the other hand, as merely any similarities of appearance not resulting from a common underlying blueprint or Platonic essence.

Darwinists have, quite reasonably, recognized that resemblances caused by a basic plan are referable to common evolutionary origin, so that the concept of homology has come to refer to commonality of ancestry. The fate of Owen's "analog," however, has been more tortuous because Owen himself could not explain the cause of resemblances not resulting from the basic plan. It seems likely that the arguments concerning analogous characters stem originally from Owen's confusion, which has been maintained to contemporary times.

APPLICATION OF THE CONCEPTS

Homology, analogy, homoplasy, and like terms represent concepts with fundamentally similar applications. Each relates two or more characters possessed by different animals, although it is convenient to discuss such relations as if only two characters are being compared at a given time. The concept defines a similarity or commonality of the two characters.

There are two elements in the specification of commonality: the kind, which is embedded in the definition of the concept employed, and the specific similarity of the character pair in question. When two characters share a kind of commonality they can be considered members of the same set: the set of all things possessing the commonality. Therefore, application of a concept requires four stipulations:

1. the first character;
2. the second character;
3. the set of commonality to which they both belong (concept); and
4. specific characteristic of the set

For instance, it may be said that a bird's wing (character) is analogous (concept) to an insect's wing (character) as a wing (characteristic); or that a bird's wing is homologous with a bat's wing as a forelimb (but not as a wing because the forelimb of the common reptilian ancestor was not a wing). It is now possible to scrutinize the basis of the concepts.

COMMONALITIES OF CHARACTERS

The arguments concerning how the concepts of homology, homoplasy, and analogy shall be defined have two causes: (1) disagreements or vagueness about how the commonalities of characters shall be assessed; and (2) disagreements about which commonalities shall be used to define a given concept. First, the three kinds of commonalities used in defining the concepts can be examined: (a) common ancestry, (b) common function, and (c) common appearance.

Common Ancestry

Common ancestry is shared by two characters if in principle each can be shown to have evolved from the same character in a common ancestor. Operationalism is more difficult to apply here than may be realized. Purists demand that the common ancestor be traceable in the fossil record, but this demand can never be met. To show that a given species (and hence any character it possesses) is ancestral to another it is necessary to have a complete genealogy of specific offspring. The fossil record may never provide all the necessary "begats" treasured by Biblical scholars. Interpolations must always be made about forms intermediate between ancestor and extant species. True, *Archaeopteryx* is intermediate between birds and ancestral reptiles, but there is no known form that stands squarely between *Archaeopteryx* and modern birds, nor between it and ancestral reptiles. Indeed, it is fortunate when paleontologists agree that a fossil form is in their best judgment a common ancestor of two extant forms. I shall call this situation a Type A judgment.

Frequently, there is no fossil of the presumed ancestor but its existence can be inferred from fossil lines before and after its time of occurrence. For example, the two principal orders of modern amphibians (Anura and Caudata) are represented by separate fossil lines in the Cenozoic but not in very early records, such as in the Carboniferous, where a single line of relevant amphibians, the Lepospondyli, occurs. By interpolation of the fossil records, paleontologists judge that somewhere a common ancestor of modern amphibians exists: possibly a Lepospondyl derivative somewhere in the Triassic or Jurassic. (Of course, the earliest tetrapods must be common ancestors, but implicitly it is the "latest" common ancestor in phylogeny that is sought.) When the common ancestor does not exist but in the best judgment of paleontologists is inferred by interpolation of the fossil record, I shall term the situation a Type B judgment.

Finally, when little or no fossil record extends backward in time from two extant forms, a common ancestor may still be inferred, because one can always be found by going back far enough (e.g., all reptiles are descended from the first tetrapods). This kind of inference is probably the most common one made by modern systematists, whose taxonomic arrangements reflect their best judgments about phylogeny. Therefore, judging from the characters of extant species, family groups of birds are erected with the implication that they share a later common ancestor among themselves than with species of other avian families. In this case, neither the common ancestor nor the fossils intermediate between it and the extant forms have been found, yet the existence of that ancestor is still credited. This situation I shall term a Type C judgment.

The implications of these practical facts are clear. If it is desired to define "common ancestry" of a character, it must always be done by inference. There exists a loose hierarchy of inferences such that Type A judgments are usually considered more reliable than Type B, and Type B more reliable than Type C. However, if an operational definition of common ancestry is to be made, it must be something such as: two characters have common ancestry if qualified experts state that they do.

Common Function

Common function of two characters is a notion that has often proved confusing. The confusion arises from ambigous uses of "function." It can be said, for instance, that a human arm "functions" as a lever, meaning it behaves according to certain rules governing levers. It can also be said that an automobile "functions" properly, meaning it performs in the way it was designed. As a result of this kind of usage, anatomists often cite "structure–function" parallels as a fundamental phenomenon in biology. It is then tempting to say that an arm of a man and the limb of an arthropod have the "same function" (i.e., that of a lever) or to state that a structure "functions" if something reasonable happens.

To the evolutionary biologist "function" definitely means something quite different and specific: it means "selective advantage" (Hailman, 1967; Klopfer & Hailman, 1967). A function of eggshell-removal behavior in nesting gulls was shown by Tinbergen, Broekhuysen, Feekes, Houghton, Kruuk, and Szulc (1962) to be antipredator in nature. Shell fragments from newly hatched eggs show the white interiors that attract crows and other predators to the nest. Tinbergen was able to demonstrate this by comparing predation rates on nests with and without eggshells.

It is not necessary for two characters to have either common ancestry or common appearance in order to be functionally similar. For instance, fox sparrows (*Passerella iliaca*) scratch in leaves to uncover insects by using a two-foot kick, presumably evolved from a hopping locomotion. Brown thrashers (*Toxostoma rufum*) turn leaves by picking them up singly with the beak and casting them aside. The two behavioral patterns are functionally similar because each is maintained by natural selection through the same forces: birds that fail to turn the leaves may die of starvation.

In theory the function of a behavioral pattern can therefore be judged only by comparing forces of natural selection on variants within a population. Theory, however, is far from practice. Tinbergen could have measured predation rates on natural gull nests and noted differences between nests where the parent did and did not show eggshell-removal behavior—that is, he could have done it if he could have found sufficient variation. All gulls he studied showed the behavior, so he had to mimic the situation experimentally by setting out his own nests with and without eggshells. The selective advantage, or biological function, of leaf-turning behavior is yet more difficult to assess, because finding and following variant individuals without the behavior is nearly impossible under natural conditions. Observations of the full range of feeding behavior in its natural context, however, provides reasonable inferences as to function. In sum, for all practical purposes it must be stated that two characters show common function if qualified experts say they do.

Common Appearance

Common appearance plays a large role in the discussion by Campbell and Hodos (1970) yet remains an elusive notion. Such workers as Boyden (1969), who has wished to use homology without phylogenetic implications, have a strong predilection to choose criteria for "common appearance" that phylogenetists consider evidence for common ancestry. For instance, the association of the radius with the ulna and their topological relations with other bones of the forelimb lead both camps to "homologize" bones of birds and mammals. The difference between the camps is that the structurists want to omit mention of phylogeny by sticking to Owne's pre-Darwinian criteria for "homolog," whereas the phylogenists want to utilize the same comparison in an evolutionary setting.

It seems clear, however, that two characters can be similar in physical appearance without necessarily sharing common ancestry, so that the notion of common appearance is a potentially useful one. For instance, the "horns" of certain rhinoceros beetles (Subfamily Dynastinae) look similar to the horns on the mammalian rhinoceros yet probably do not serve the same function and certainly are not derived from "horns" of any common ancestor. Naturally, anatomical investigation shows that the two kinds of "horns" really bear little resemblance to one another, yet as "horns" on the front of the animal they do look similar. It is doubtful that any operational definition of common appearance can ever be agreed on, yet ironically the notion remains useful in describing characters of animals. It must therefore be concluded that two characters have common appearance if anyone says they do.

DEFINING CONCEPTS BY VENN DIAGRAMS

The arguments (see Campbell & Hodos, 1970) as to how concepts of homology, analogy, and homoplasy should be defined all revolve around various combinations of the three kinds of commonalities discussed above. It is therefore useful to employ Venn diagrams to show clearly the logical relationships among these definitions. In Figure 1 the three circles represent the sets of character pairs that have the three commonalities: common ancestry, function, and appearance. These three sets define eight topological areas, and every character pair must fall

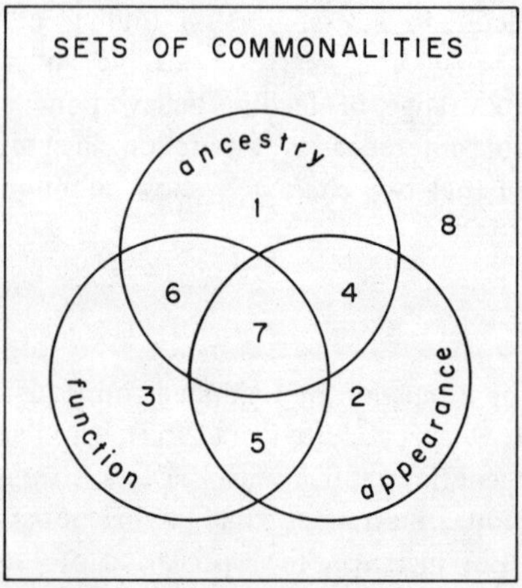

FIG. 1 A Venn diagram of logical relations among three sets of commonalities that may be possessed by a pair of morphological or behavioral characteristics. A given pair of characters may fall into one of eight topologically defined logical areas of the diagram, and there are 256 different ways to shade such a diagram in order to define logical inclusiveness of terms such as "homology." Examples of such shading are given in Figures 2 and 3.

in one of the eight areas. By scrutinization of the definitions offered for the concepts of interest, the Venn diagram can be shaded according to which of the eight areas is included in a particular author's definition. It is difficult to believe that there are 256 different ways to shade Figure 1. There are three routes for doubting Thomases. First, if the number of shaded topological areas is i, then the number of ways to shade i areas among the eight total is eight things taken i at a time: $\binom{8}{i}$. Therefore, the total number of ways to shade is

$$\sum_{i=1}^{8} \binom{8}{i} = 256.$$

Second, each area has two possible states, shaded and unshaded, and there are eight areas so that $2^8 = 256$. The third proof is to provide the 256 different diagrams, which I shall be happy to do for any editor who promises to print them! Happily, the controversy over concepts of homology and its sister concepts has generated only a fraction of the 256 possibilities, which may now be considered in turn.

DEFINITIONS

Campbell and Hodos (1970, p. 358) define homology thus: "structures and other entities are homologous when they could, in principle, be traced back through a genealogical series to a stipulated common ancestral precursor irrespective of morphological similarity." This definition is represented in Figure 2a. It seems clear that only areas within the "common ancestry" circle can be shaded and that morphological similarity (common appearance) is irrelevant, so that Areas 4 and 7 are to be shaded. Because functional similarity is not

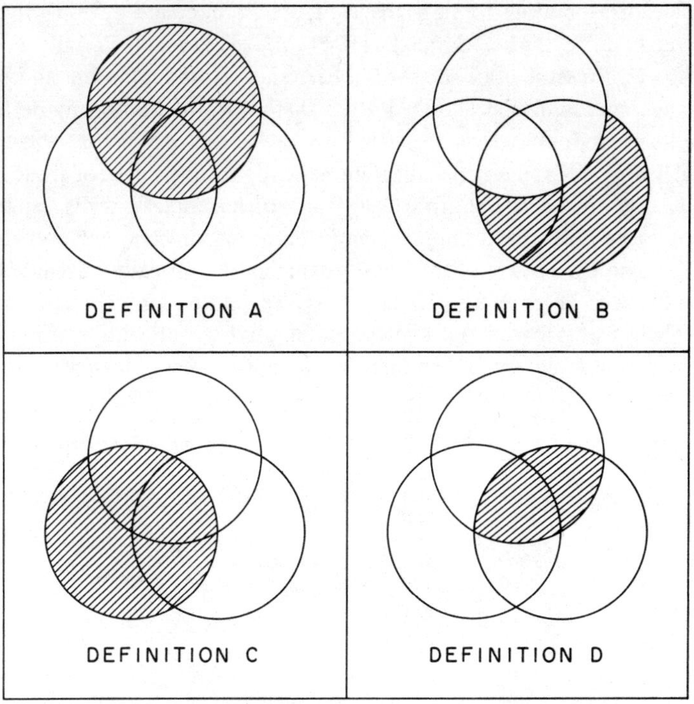

FIG. 2 Logical shadings of the Venn diagram of Figure 1 to show definitions of concepts offered by Campbell and Hodos (1970). Definition A is their "homology," Definition B "homoplasy," Definition C "analogy" and Definition D "homogeny." Other authors define these same terms differently and use these definitions for other terms. For example, Simpson's (1961) "homology" appears to be that of logical Definition D. Also see Figure 3.

mentioned, it can be assumed that it is also irrelevant and shading can be done accordingly.

The same authors (Campbell & Hodos, 1970, p. 359) define homoplasy as "structures or entities that are morphologically similar but that cannot, in principle, be traced back to a stipulated common precursor." Again, functional similarity is omitted and can therefore be assumed irrelevant, so that Figure 2b is shaded accordingly. Last, analogy is defined as "correspondence between structures or entities due to similarity in function whether or not they can be traced to a stipulated common precursor." Figure 2c shades the Venn diagram accordingly.

It should also be noted that Campbell and Hodos (1970, p. 359) define a fourth concept heretofore not mentioned: "homologous structures that are also morphologically similar are homogeneous." The definition of homogeny is shown diagramatically in Figure 2d.

Without association of the terms of Campbell's and Hodos' (1970) terms to these definitions in Figure 2, they may be referred to merely as Definitions A, B,

C, and D (corresponding to diagrams in Figure 2a, b, c, and d, respectively). It is interesting to note that if Simpson (1961, p. 78) is taken literally in defining homology as "resemblance due to inheritance from common ancestry" his concept becomes equivalent to Definition D (that is, the homogeny definition of Campbell and Hodos).

Figure 3 presents some other diagrammatically defined uses of these terms. In Figure 3e (Definition E), Smith's (1967) use of homoplasy is diagrammed; this usage embraces both the homoplasy and homogeny of Campbell and Hodos (cf. Figure 2, b and d). To complicate matters further, historically misguided authors use Definition E (Figure 3e) for homology (e.g., Boyden, 1969; Kuhlenbeck, 1967). Their use is based on a rejection of phylogenetic implications and by default relies completely on similarities of appearance. However, the criteria

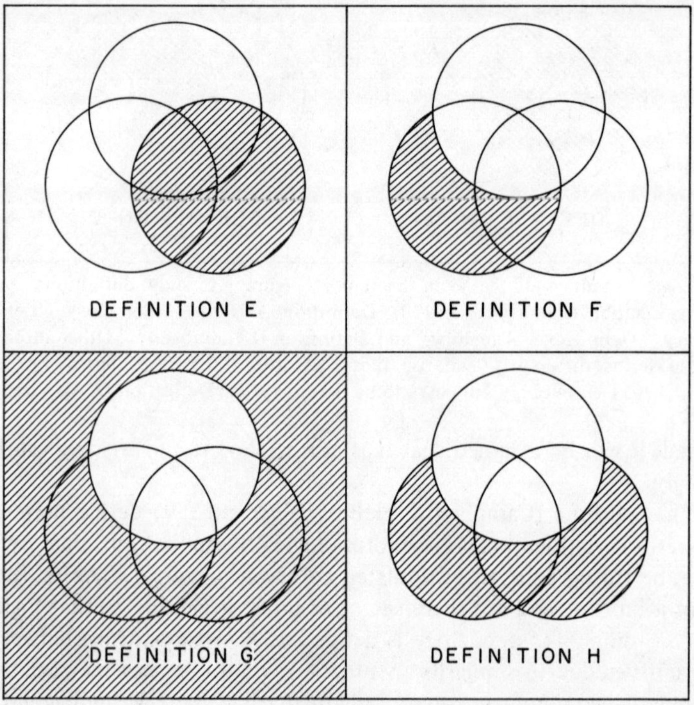

FIG. 3 Other logical shadings of the Venn diagram of Figure 1, which show definitions of concepts by various authors (Cf. figure 2). Definition E is Smith's (1967) "homoplasy," logically equivalent to Definitions B plus D in Figure 2. Other authors use Definition E for "homology" (see text). Definitions F and G have been used to define "analogy" by modern authors, whereas historical evidence suggests that Owen's original intent may have been Definition H. These definitions of Figure 2 and 3 are used in Table 1 to review meanings of concepts used by different authors.

they use to define a homologous similarity are very similar to those used by phylogenists to infer common ancestry (a point that seems to have been bypassed in the discussion by Campbell & Hodos, 1970). Ironically, then, despite differences in logical content of their stipulated definitions, the modern structuralists and phylogenists often designate the same character pairs as homologous.

Cracraft (1967) has reviewed the uses of the term "analogy," showing that it has been used in at least three different ways. One of these is Definition C, as used by Campbell and Hodos (1970). Another (Figure 3f) excludes characters that share ancestry in addition to function, whereas a third definition of "analogy" (Figure 3g), derived from Bock (1963), treats all nonhomologous characters as analogous. This last approach would again be a misreading of Owen, who would not have considered characters having nothing in common as analogous. Because Owen had no real idea of selective advantage or biological function in a pre-Darwinian era, he could not clearly have made the distinction between common function and common appearance. However, Ghiselin (Chapter 9 in this volume) has shown clearly that Owen practices a belief that "analogues" are characters sharing appearance not resulting from homology (Definition B) despite his lexical definition of functional similarity (Definitions C and F). Indeed, the fairest statement of Owen's conceptual thinking is probably represented by Figure 3h, although again it is worth emphasizing that reading a Darwinian interpretation into a Platonic one is always doomed to intellectual failure.

The definitions represented by Venn diagrams A through H are not necessarily a complete list. Nevertheless, terms (homogeny, homology, homoplasy, etc.) can be compared with logical contents (Definitions A through H) to summarize some of the diversity of useage in Table 1. It is quite clear from the table that a given term has different logical contents in the minds of different authors and that a given logical content has been represented by different terms. The question is not to decide future usage by historical precedent but to decide how many terms are actually required and then to assign and stabilize their meanings.

DEFINITIONS AND INFORMATION

Criteria for Efficient Definitions

To be useful, terms must communicate unambiguously. So long as what Campbell, Hodos, Bock, Ghiselin, Simpson, Boyden, Kuhlenbeck or Owen mean by "homology" is understood correctly it is irrelevant that they all use the same meaning. Yet it is certainly convenient if they do. So if a *stable* meaning for a concept is sought for convenience in communication, *efficiency* might also be

TABLE 1
Partial Survey of Explicit Definitions, Implied Meanings, and Apparent Uses of Terms Employed or Discussed by Various Authors[a]

Closest logical equivalent (Figs. 2 & 3)	Term			
	Homology	Analogy	Homoplasy	Homogeny
A	Owen (1843) Bock (1963) Ghiselin (1966) Campbell and Hodos (1970)			
B		Boyden (1969)	Simpson (1961) Campbell and Hodos (1970)	
C		Owen (1843) Cracraft (1967) Campbell and Hodos (1970)		
D	Simpson (1961)			Lankester (1870) Campbell and Hodos (1970)
E	Boyden (1969) Kuhlenbeck (1967)		Smith (1967)	
F		Cracraft (1967)		
G		Bock (1963) Cracraft (1967)		
H		Owen (1843)		

[a]This table is a guide to some relevant literature, intended to demonstrate the diversity of meaning applied to various terms. The table does not imply than an author necessarily endorses the usage listed for his name but merely that he at least discusses such usage.

required. Although there is no objective way to decide which definition of homology is "best," there is a way to judge what sort of system of definitions communicates most efficiently.

There appear to be no empty sets among the eight topological areas in Figure 1, although I shall avoid here the labor of providing examples for each. Suppose that character pairs of interest have an equal probability of falling in any of these eight areas. Under such conditions, specification of a unique area for a given character pair communicates exactly three bits of information, in the technical sense of information theory (Shannon & Weaver, 1963). A numerical system for such specifications could use the numbers of Figure 1, so that saying "character pair *X* is in area 6" communicates three bits of information.

Suppose, under the same conditions of equiprobability, it is desirable to use a linguistic system in place of the number designation. For instance, each numbered area in Figure 1 can be given a unique name, such that each name then carries the three bits of information. Alternatively, names could be used to represent groups of areas in Figure 1 and a given area specified by some combination of names, noting whether the name did or did not apply to the character pair of interest. For instance, the "ancestry," "appearance," and "function" circles of commonalities in Figure 1 could be selected as names. Three bits of information can then be communicated with a statement such as "the character pair shows common ancestry and appearance but not common function."

A fourth approach to the problem of communication is to define certain words as including certain areas that are not necessarily coincident with the three circles of Figure 1. This is essentially the approach represented in Definitions A through H illustrated in Figures 2 and 3. We then communicate by stating whether or not such concepts apply. For example, take Definition A (the "homology" of Campbell & Hodos, 1970), Definition C (the "analogy" of Campbell & Hodos, 1970), and Definition E (the "homoplasy" of Smith, 1967). These definitions can be coded with the letter if the definition applies and with a barred letter if the definition does not apply. Therefore, to designate a character pair in Area 6 of Figure 1, "A, C, and E" (the last being read "not-E") can be used. By such designation the three bits of information are communicated. Alternatively, it can be stated that the character pair shows "homology and analogy, but not homoplasy," so long as these linguistic codes are unambiguously known to all recipients of the information. If the recipient believes that the "homoplasy" of Campbell and Hodos (1970) is meant, then a message different from that intended is received.

No more than three terms are necessary for this kind of linguistic communication, as long as they are efficiently defined in terms of Figure 1. However, suppose the Campbell and Hodos (1970) definition of "homoplasy," which is Definition B, is used. A character pair in Area 6 of Figure 1 is now "A, C, and $\bar{B}$" under this new system of definitions, but a character pair in Area 7 has this designation also. Therefore, it becomes necessary to include the Campbell-Hodos concept of "homogeny" (Definition D) and to designate a character pair in Area 6 as showing "A, C, $\overline{B}$, and $\overline{D}$"–or "homologous and analogous, but neither homoplastic or homogenous." In other words, the Campbell-Hodos system can be shown to require four words to communicate three bits of information, when other systems can be shown to require only three words.

Complications of Efficient Communication

The foregoing analysis is predicated on the assumption that character pairs of interest are distributed equiprobably among the eight areas of Figure 1. This assumption is undoubtedly wrong. For instance, if a pair of characters has nothing in common (Area 8), then it is quite unlikely to be the subject of scientific discussion. In order to devise the most efficient linguistic coding system possible, it is necessary to know the probabilities associated with each topological area in Figure 1. Because these are not known (at least to me), I can offer no suggestions for a final scheme. I have elsewhere (Hailman, 1973) shown how ornithologists have efficiently handled a similar problem with an eight-area logical space, and a parallel kind of system may be applicable here.

If efficiency of communication is to be taken as a criterion for how concepts shall be defined, other complications require attention. For instance, it is not immediately evident what shall be considered "efficient." There are, at maximum, three bits of information to communicate–less than that if the eight areas in Figure 1 are not equiprobably filled by character pairs of interest. If communication is with one column of a computer card, assigning a digit (or letter) to each of the eight areas (as in Figure 1) is most efficient–the system requires only one column. In contrast, if communication is via speech or the printed word, number designations may prove cumbersome because they are not immediately recognized. The message is transmitted efficiently but its encoding and decoding may require reference to a diagram, such as Figure 1, in which the number is translated out of and back into linguistic terms. Therefore, in ordinary human communication among scientists, the encoding–decoding process may prove grossly inefficient, both in terms of time required to communicate and in fidelity of getting the intended message through. I therefore opt for linguistic codes.

Even with linguistic codes, efficiency of communication may be difficult to define. If a given character pair shows "homology and analogy but not homoplasy" (under someone's agreed on definitions), efficiency can probably be improved by dropping the "not homoplasy" and leaving it to be understood by omission. This shortens the linguistic designation but at the same time leaves room for unintentional omissions that garble the message. Moreover, as any editor can testify, the shorter the word, the better for journals and their limited space. Therefore, there is yet another sense of "efficiency" that favors shortening such words as "homology" to "hogy" and "homoplasy" to "hasy"–or something equally barbarous.

Finally, there is the consideration so dramatically illustrated by Table 1, which implies that efficiency should take into account the confusion of multiple past usages of the same word. This kind of efficiency eliminates from the new vocabulary any term that has had more than one meaning in the past–which means eliminating such words as "homology" and "analogy."

Information and Efficiency

So far as I am aware, the application of information theory to efficiency of scientific concepts (Hailman, 1973) is a new suggestion. The foregoing discussion has explored the difficulties of this approach at length not so much for purposes of devising the ultimate scheme of treating "homology" problems as for explicating in a complete framework the nature of considerations that should go into scientific definitions. Usually, debates among scientists as to the definitions of terms are fought on grounds considering only such things as stability of usage, historical origins, and linguistic appropriateness of words chosen. Insufficient attention is paid to the logical content of words, the amount of information they convey, and the efficiency with which information is transmitted. As I cannot offer an ultimate solution to the "homology debate," I have tried to do the next best thing: to articulate and emphasize the important issues to be discussed in the debate.

OTHER PROBLEMS IN HOMOLOGY

For the remainder of this contribution, I shall use "homology" to represent "common ancestry" (Definition A in Figure 2) without apology, unless the context shows I am indirectly quoting other uses of the term.

Behavioral Homology

Hodos (Chapter 11 of this volume) has shown that the decision as to whether two behavioral characters share a common ancestry is in practice a difficult one, and in this sense he agrees with Atz (1970). One continuing point of misunderstanding between behavioral scientists, who believe that behavioral patterns can be homologized, and morphologically oriented scientists, who dwell on the difficulties of such a task, lies in the idea of "function," which is discussed above. Both Hodos and Atz treat behavior as the "function" of structure and find it difficult to consider behavior apart from the structure that underlies it.

There are two points that seem to require more emphasis in the attitudes that structurally oriented scientists have toward the evidence of behaviorists for homology. The first concerns the "minuteness of similarity" criterion frequently applied to morphology, in which pervasive point-to-point similarities between two structures are sought as evidence for homology. Behavior is often implicitly referred to a more vague level of accessibility, in which such minutenesses of similarity are impossible to establish. There seems, however, to be no real philosophical distinction here, for in theory at least the details of behavior are as accessible to measurement as are those of morphology. Atz (1970) seems to recognize this implicitly when he accepts homologizing of fixed action patterns (which have been measured in admirable detail in some cases) as "more valid" than other kinds of behavioral homologies.

The second criterion used by morphologists is the topographic relations among parts: what the character in question is connected to and where it lies in relation to other characters. Atz (1970) refers to this as the "criterion of position" but is apparently unaware that a kind of "position" exists in behavior as well as morphology. Many action patterns are found to occur in temporally ordered, spontaneous, output strings. Indeed, the analysis of such temporal strings is beginning to fill the ethological journals and although the "criterion of temporal position" does not yet appear to have been used in homologizing behavior, it seems inevitable that it will be.

Lest misunderstanding arise, I emphasize that I am not asserting that behavioral patterns are "just like anatomical structures." For one thing, behavior has a kind of variability unknown in morphology. The length of a bone can be measured repeatedly and a frequency distribution of measurements obtained that relate to inherent problems of measurement. The same problems apply to measuring some variable in the motion picture of a fixed action pattern. However, if the same variable is measured in repeated performances of the "same" fixed action pattern of a given individual animal, the variability in the measurements increases because there is a source attributable to differences among repeated performances (e.g., Hailman, 1967).

In sum, it can be noted that homologizing of behavior, especially fixed action patterns, has been a fundamental interest of ethologists dating from the founding of the young science by Lorenz, Tinbergen, and a few others. It is hoped that the criticisms of structurally oriented scientists can prove useful in prodding ethologists to deliver their criteria more explicitly, but no amount of criticism can halt the use of homology concepts in behavior. The attitude shown by Atz (1970) and Hodos (Chapter 11 of this volume) that homologizing of behavior must await exploration of the structures underlying the behavior, and then their homologizing, is not likely to gain favor because it is so unrealistic. Atz accepts behavioral homologies of fixed action patterns among closely related species but puts a value judgment on other comparisons, which he calls "nontrite" (Atz, 1970, p. 69). It is not clear why Atz (1970, p. 56) thinks measurement of behavior is "like whistling in the dark," but I expect that most ethologists will continue to whistle.

Kinds of Homology and Analogy

Ghiselin (Chapter 9 of this volume) has wrestled with some of the minor problems concerning kinds of homology and analogy that arise in animal structures. (His conception of "analogy" seems to be close to that of Definition H above.) Of interest for final comment here is the problem of behavioral ritualization vis-à-vis Ghiselin's "iterative homology," which he defines as entities traceable "to the same element of a precursor which causes the repetition of parts in a larger whole." Classical serial homology is one of the subsets included.

There is a special problem in behavioral analysis that seems to demand attention. Ethologists believe that most of the communicatory patterns of animals are evolved from more basic noncommunicatory patterns, a process that Tinbergen (1948, 1952) has called "ritualization." The result of ritualization in the behavior of an extant animal is that both the unritualized and the ritualized forms may occur in the same animal. For instance, Andrew (1956) studied the tail-flicking movements in a large sample of avian species, finding that the evident similarities that occurred correlated with taxonomic boundaries, i.e., a classical behavioral homology and one not cited by Atz (1970). However, within a species, these simple incipient movements of flight may be ritualized into signals for communication. To give a dramatic example from my own studies, the dark-eyed junco (*Junco hyemalis*) flicks its tail feathers open when feeding alone, making it quite visible to conspecifics flying above—a signal that appears to help promote social grouping. In this case, the outer tail feathers have become white, contrasting sharply with the slate gray of the rest of the tail and supplying a coloration that emphasizes the signaling movement of tail flicking. The point is that there is present in the same species (in the same individual to be precise) the basic unritualized tail flick of incipient flight and the derived, ritualized tail flick. Are such cases of what ethologists may consider "obvious" commonality of ancestry to be considered as an example of a yet different kind of "iterative homology"?

Finally, as an attempt to integrate the problems of analogy raised by Ghiselin (Chapter 9 of this volume) and the viewpoint taken here, I attempt to relate his subsets of analogy to the diagram of Figure 1. "Convergent analogy"—characters that share appearance because of their evolution by similar selection pressures—clearly belongs in topological Area 5. "Parallel analogy" is a difficult concept and one that has always proved troublesome, for it implies that two structures have evolved similarly both because they share a common ancestor and because they have experienced similar selection pressures. Parallel evolution of characters therefore lies somewhere between true homology (in the sense of Definition A) and convergent analogy just discussed, for the common ancestor possesses only the evolutionary basis for the independently derived similarities. Parallel evolution of characters therefore does not fit nicely into the scheme of Figure 1, but it can also be argued that it does not fit nicely into anyone's modern conception of analogy either. Ghiselin (Chapter 9 of this volume) is able to call the case a subset of analogy only by defining analogy in a way that ignores the distinction between commonalities of appearance and function. It seems to me that Ghiselin has again served clarity of scientific communication by focusing on a topic of interest that deserves more thoughtful attention.

Ghiselin's "coincidental analogy" clearly belongs in Area 2 of Figure 1, where commonalities of appearance are caused by neither common ancestry nor common function. His "iconic analogy," which most investigators refer to as mimicry, is another kind of concept that usually belongs in Area 2 of Figure 1, because in many cases the selection pressures producing the mimic differ from those producing the model (i.e., Ghiselin's example of a leaflike insect resembling a tree leaf). However, a viceroy butterfly and a monarch are brightly colored aposematically, natural selection removing deviants that do not display the warning coloration. Should such cases, in which there is still an evident difference in the two species, be considered to have the "same function" (i.e., be placed in Area 3 of Figure 1)?

In one sense, the warning coloration of the distasteful model evolved under the influence of multiple selection pressures: aposematic coloration differs greatly among species and seems to have only the necessity of being visually distinctive and usually highly colored. Therefore, other considerations must have determined the exact *kind* of aposematic coloration assumed by the distasteful monarch. The mimicking viceroy, however, could assume only the coloration of its monarch model to be maximally effective, so that the selection pressures of predation can account entirely for the evolution of its coloration. It seems that once again Ghiselin has fingered a conceptual problem that present terminology is inadequate to handle, although again his own treatment so far fails to tease apart considerations of common appearance and function in his conception of analogy.

SUMMARY AND OVERVIEW

Such terms as "homology" and "analogy" refer to various sorts of commonalities of characters. Their origins, as pointed out by Ghiselin in this volume and elsewhere, are pre-Darwinian and possibly Aristotelian in origin. The establishment of evolutionary thinking led to a divergence of interpretation, phylogenists emphasizing that the common blueprint of homologous structures was the ancestral character of modern homologous derivatives and the modern structuralists emphasizing the morphological operationalism of the homology concept. Because the latter tend to choose criteria for morphological correspondence that the former accept as evidence for common ancestry, both schools tend to apply "homology" in similar ways.

Other kinds of difficulties concern the definitions of such terms as "analogy" and "homoplasy," which are defined variously by different authors. I have attempted here to reduce the logical content of all these terms to their inclusion or noninclusion of three kinds of commonalities that may be identified among characters: common ancestry, common function, and common appearance (Figure 1). A large number of possible concepts is definable by these three commonalities (Figures 2 and 3) and various definitions have been used for the same term (Table 1).

If meanings for terms are now to be stabilized, it seems desirable to make new definitions efficient in communicating information among scientists. A criterion of efficiency, derived from the mathematical theory of information, has been presented and difficulties attendant to its application have been discussed. Whatever final agreement is reached, it should be structured so as to emphasize the logic, informational content, and efficiency of scientific communication rather than the historical origins of stability per se.

Difficulties of homologizing behavioral patterns raised by Atz and Hodos are viewed as problems of application and evidence rather than as fundamental logical problems. Ghiselin's broadening of the concept of serial homology is extended further to the problem of the ritualization of behavioral patterns and other subsidiary issues raised by Ghiselin are placed in the framework of Figure 1. Parallel evolution or analogy emerges as a difficult logical problem, as does the treatment of similarities resulting from classical Batesian mimicry.

I suggest in conclusion that certain terms relating to commonalities of characters may have outlived their usefulness to precise science. Such terms as "instinct" (Hailman, 1967, 1969) and "drive" (Hinde, 1959, 1970) remain in ethology as identifications of problem areas, not as specific concepts, and it seems that "homology" and its kin may be destined for a similar fate.

REFERENCES

Andrew, R. J. Intention movements of flight in certain passerines, and their use in systematics. *Behaviour,* 1956, **10,** 179–204.

Atz, J. W. The application of the idea of homology to behavior. In L. R. Aronson, E. Tobach, D. S. Lehrman, & J. S. Rosenblatt (Eds.), *Development and evolution of behavior.* San Francisco: W. H. Freeman, 1970. Pp. 53–74.

Bock, W. J. Evolution and phylogeny in morphologically uniform groups. *American Naturalist,* 1963, 97, 265–285.

Bock, W. J. The concept of homology: Discussion. In Peters and Noback, Comparative and evolutionary aspects of the vertebrate central nervous system. *Annals of the New York Academy of Science,* 1969, **167,** 71–73.

Boyden, A. Homology and analogy. *Science,* **1969, 164,** 455–456.

Campbell, C. B. D., & Hodos, W. The concept of homology and the evolution of the nervous system. *Brain, Behavior, & Evolution,* 1970, **3,** 353–367.

Cracraft, J. Comments on homology and analogy. *Systematic Zoology,* 1967, **16,** 355–359.

Ghiselin, M. T. An application of the theory of definitions to systematic principles. *Systematic Zoology,* 1966, **15,** 127–130.

Ghiselin, M. T. The distinction between similarity and homology. *Systematic Zoology,* 1969, **18,** 148–149.

Hailman, J. P. Ontogeny of an instinct. *Behaviour Supplement,* 1967, **15,** 1–159.

Hailman, J. P. Instinct. *Encyclopedia Americana,* 1969, **15,** 223–224.

Hailman, J. P. Ecoethological aspects of reproduction: Discussion. In D. S. Farner (Ed.), *Breeding biology of birds.* Washington, D.C.: National Academy of Sciences, 1973. Pp. 22–27.

Hinde, R. A. Unitary drives. *Animal Behavior,* 1959, **7,** 130–141.

Hinde, R. A. *Animal Behaviour.* (2nd ed.) New York: McGraw Hill, 1970.

Klopfer, P. H., & Hailman, J. P. *An introduction to animal behavior.* Englewood Cliffs, N.J.: Prentice-Hall, 1967.

Kuhlenbeck, H. *The central nervous system of vertebrates, I. Propaedeutics to comparative neurology.* New York: Academic Press, 1967.

Lankester, E. R. On the use of the term homology in modern zoology, and the distinction between homogenetic and homoplastic agreements. *Annual Magazine of Natural History,* 1870, **6,** 34–43 (not seen).

Owen, R. *Lectures of the comparative anatomy and physiology of the invertebrate animals.* London: Longmans, Brown, Green, and Longmans, 1843, (not seen).

Shannon, C. E., & Weaver, W. *The mathematical theory of communication.* Urbana: University of Illinois Press, 1963.

Simpson, G. G. *Principles of animal taxonomy.* New York: Columbia University Press, 1961.

Smith, H. M. Biological similarities and homologies. *Systematic Zoology,* 1967, **16,** 101–102.

Tinbergen, N. Social releasers and the experimental method required for their study. *Wilson Bulletin,* 1948, **60,** 6–51.

Tinbergen, N. Derived activities: Their causation, biological significance, origin and emancipation during evolution. *Quarterly Review of Biology,* 1952, **27,** 1–32.

Tinbergen, N., Broekhuysen, G. J., Feekes, V., Houghton, J. C. W., Kruuk, H., & Szule, E. Eggshell removal by the black-headed gull, *Larus ridibundus L.,* a behaviour component of camouflage. *Behaviour,* 1962, **19,** 74–117.

15

Problems in the Comparative Study of Language

Eric H. Lenneberg

Cornell University

The motivation for including language among the topics of this volume is, undoubtedly, the hope of seeing man in the context of his evolution from the lower primates. Similarly there is, perhaps, the hope that it may be possible to relate language to brain function in man and therefore to see how the human brain evolved. These are realistic expectations in principle but in practice there are difficulties.

THE PROBLEM OF HOMOLOGIES WITH RESPECT TO LANGUAGE

The difficulties in homologizing behavior have been discussed repeatedly in the past (Roe & Simpson, 1958; Atz, 1970; Lenneberg, 1969) and need be reviewed here only in briefest form. I should like to concentrate on the notion that language is specific to the human species. Unfortunately, this issue has frequently been confused with another one, namely that language is innate. From the outset, let it be clear that these are quite independent issues–one can maintain the notion of species specificity without at the same time adhering to a nativistic view of language. Although I have used the word "innate" myself in the past, I have always been careful to define it in my own terms, and I should like to state here unequivocally that I do not wish to reintroduce the eighteenth century doctrine of innate ideas. The concept of species specificity has nothing to do with the factors responsible for calling forth or shaping a particular type of behavior (such as language) in the course of ontogenetic development. The only question to be considered here is whether species specificity is compatible with modern theories of evolution.

There are both theoretical and empirical reasons that species specificity is not merely compatible with evolutionary theory but, in fact, is to be expected from what is known about phylogeny. In the first place, it is well established that a species may undergo evolutionary changes *in toto*. Not all changes in the course of evolution are accompanied by branching. All that is necessary for evolutionary changes to take place without branching is for a population to have an opportunity for active gene exchange. Even a cursory survey of the animal kingdom makes it plain that some species reached plateaus of stability long ago; their morphology and function have undergone only relatively minor changes during the past many millions of years. However, especially among the so-called "higher" animals, instabilities are more common; perhaps the more complex the animal and the longer its embryogenesis, the greater becomes the likelihood of polymorphism and deviations. Whether or not branching occurs is probably more a function of environmental conditions, geographic distribution, and population mechanical parameters than it is of inherent factors in the genetic peculiarities of an organism. Therefore, the simple fact that species are quite likely to undergo changes as a whole without branching is sufficient to account for species specificities. It is true that sibling species, races, or forms may be found so closely related that it must be assumed that the species is in the process of branching; in these cases, morphological, physiological, and behavioral features will be found that seem to grade imperceptibly from one form to its cognate. Much more commonly, however, species have some distinctive features. Individual differences do, of course, occur within the species, but this variability does not usually bridge the gap between related species. This is not to deny that common denominators may also exist between taxa. Even in the realm of commonalities, however, the species's own signature may frequently be discerned. My first point, therefore, is that there is nothing unusual about the assertion that man's behavior may have certain species specificities. It is entirely compatible with a Darwinian orientation. Second, I do not deny that commonalities exist in the area of primate behavior; however, I expect, if for no other than theoretical reasons, that each primate species also has species-specific features in its behavioral traits. Because the common origin of man and ape lies in the fairly distant past, it is not surprising that their social organization and the form and content of their communicative behavior are definitely distinct (qualitatively and quantitatively).

It is also widely agreed that evolutionary change takes place in small steps and that what changes is the molecular substrate of inheritance. There are two reasons to believe that even small molecular changes may at times result in rather dramatic differences in the population of mature individuals of a species. In the first place, there are examples in which individuals suffer small genotypic changes with only minimal phenotypic alterations, i.e., where the developmental end product, the adult phenotype, may remain quite constant in the face of rather considerable variation in the genome. This situation seems to set the stage for apparently sudden phenotypic jumplike changes; the genome is yet further altered only minimally but a threshold of tolerance for phenotypic constancy is crossed, so that a rather distinct new phenotype emerges (the straw that broke the camel's back phenomenon), discussed in the literature under the heading of canalization (see especially Waddington, 1967–1970).

In the second place, it is known that some minimal molecular alterations in the genome may have far-reaching consequences for the organism's development, be amplified so to speak, during embryogenesis to produce phenotypes with properties quite different from those of a given standard genome. Because both of these conditions may be found among many species, there is no strong reason to believe that they have not at times played a role in the process of speciation. Species have evolved and are evolving at sharply varying rates. This is simply to say once more that there need be no worry about finding certain peculiarities in man not shared with him by any other primate. Any assertion to this effect must stand the proof of empirical verification but is, in itself, not an antievolutionary concept.

Thus far, I have discussed merely differences and discontinuities on the surface or among the end products of ontogenies. I have said nothing about the histories of discontinuities. Actually, biology is or ought to be the most history-conscious discipline among the natural sciences. In a sense, all of biology deals with the problem of change over time and of how structures and conditions come about. It is a general axiom of biology that whatever pattern or peculiarity is encountered must have come about by some transformational history—it had an origin. Many biologists try to trace these origins further and further back; their hope is to be able to reduce whatever they see to conditions of matter, that is, physics and chemistry. Some biologists scorn this form of extreme and idealized reductionism but they are, nevertheless, concerned with short-term histories. In short, the concept of species specificity does not imply a doctrine of special creation, as is frequently imputed to it. Species specificities have as much a history as species commonalities. Moreover, species specificities might very well have arisen from a stage that constituted a common origin of two subsequently diverging trends of development. In many cases, however, the protoform that has given rise to two different species may have existed so far back, and the two forms since then may have each undergone so many changes, that it becomes extremely difficult to demonstrate homologies in one species for every single trait observed in the other. According to modern usage in evolutionary theory, the word "homology" should be reserved for corresponding structures or traits that have a common phylogenetic origin. Demonstration of such a common origin is in some cases easier than in others, and there is frequently an element of speculation involved. Homologizing bones is fairly easy; muscles are more apt to pose problems; physiological processes vary a great deal in this respect. In some instances there can be no question whatever (e.g., respiration), whereas in others the phylogenetic origin is in doubt. Behavior is more difficult still. There are many species-specific behavioral traits throughout the animal kingdom that appear to have no homologies whatever. The cross-species comparison of behavior also presents another problem for the reconstruction of evolutionary origins, one that is particularly troublesome in the case of human language. Often sets of behavior are found that seem to have the same function for different animals and that seem to accomplish the same ends—and yet they are executed by totally different means involving different muscles or appendages, different motor patterns, different endogenous or exogenous stimuli, etc. When this is the case (as it is, for example, with communication between bees and communication between humans) homologies, of course, cannot be spoken of at all—at best, convergence can be considered and even this notion is often idle anthropomorphic interpretation. This raises the general question, when is it possible to speak of and search for homologies, i.e., common phylogenetic origins? Without wanting to set up doctrinaire prescriptions, I may at least assert this one indispensable prerequisite: before homologies of a trait can be sought, there must be available a reasonable definition of the trait—there must be a clear idea of what the trait is for which homologies are being sought.

When it comes to human language, rather unexpected difficulties are encountered in this respect. At first glance, the nature of language appears intuitively obvious. It looks as if it were a set of arbitrary associations between objects and symbols used for the purpose of communication. The inadequacy of such a notion, widely accepted until the late 1950s, has now been amply demonstrated and needs no further comment. Subsequently, language came to be regarded as a natural formalism that could be discovered by empirical research and described in purely formal terms; in this version (Chomsky, 1957), the natural formalism was thought to be related to objects of the world (or thoughts) in much the same way as arithmetic is related to unspecified objects; and in much the same way that common arithmetic can be described by a more abstract formalism, an algebra of some sort, so, it was thought, an abstract formalism could be discovered that would describe a natural language adequately. This early notion of *pure* formalization was soon abandoned for another descriptive system, one that incorporated semantic aspects into the general theory of syntax (Chomsky, 1965). Even today, most linguists insist that the nature of language may be accurately characterized as a set of rules that, in the ideal, constitute an algorithm for describing any sentence of a language. Supposedly, semantics is also part of the formalism and is separable from human knowledge in general.

Unfortunately, the story does not end here. It has become increasingly clear, and I personally insist on it, that the separation of human language from general human knowledge is in all but a few marginal instances impossible. Substantiations of this claim follow below. This realization is an important setback for linguistic theory, because it dashes any hope of analyzing the nature of natural languages with the degree of precision and completeness that Chomsky's early beginnings seemed to promise. In order to reach decisions on most of the significant issues in linguistic analysis and in the practical use of language, one must rely on one's own knowledge of the world. A natural language is therefore at best a semiformalism. It is an open system, partially amorphous even though it has many formal properties; and even its formal properties are frequently so imbued with unformalizable, epistemological aspects that precise and abstract description becomes just as difficult as the precise description of knowledge as a whole.

These remarks are relevant to the topic of cross-species comparison of languagelike behavior and the search for homologies because they lead to the predicament of ignorance with respect to the nature of human language. Just what is it that is to be homologized? The simple notion of communication is clearly insufficient. Inasmuch as both Premack and the Gardners have abandoned the idea of looking for language capacities in the acoustic medium, they can agree that what matters is a more abstract cognitive property of language. (There is no disagreement here.) After all, there are human beings of whom it can be said that they have language even though they can neither speak, nor hear, nor even see. They learn to use a tactual system and by this means contrive to understand and to use meaningful and acceptable sentences. (At any one time the Perkins Institute for the Blind in Watertown, Massachusetts, can be visited and such individuals found.) Those who learn to use vocal communication usually do so after the principles of language have been inculcated through the vehicle of tactual alphabets. The situation is slightly different in deaf families who use sign language. There is a time when their deaf infants are clearly without any type of language; as they grow up, however, there can be no doubt that they do have language, in both its written and its gestural form. However, during the course of development, communication may be observed between children and adults that ranges from barely languagelike to almost languagelike; for instance, tugging on sleeves, acting out in pantomime, or using stylized "ideographs." Especially the congenitally deaf illustrate how difficult it is to decide what is and what is not language. At all times it is a curious mix of knowing how and knowing that. In other words, I am finding it increasingly difficult to make a distinction between learning how to speak and learning what to say. In order to find out whether apes have language, it therefore seems necessary to me to discover not only whether they have a symbol for banana, but whether they deal with the world in just such ways as humans do when they receive and produce utterances. In principle, it should be possible to make cross-species comparisons between types of cognition, and many participants at the conference on which this volume is based have attempted to do so. However, it is obvious that there are enormous problems in making objective comparisons between man's cognition as a whole and that of other primates. What is most puzzling here is that, on the one hand, all the documented cognitive achievements of monkeys and apes are quantitatively, if not qualitatively, so distant from man's—and yet, on the other hand, that chimpanzees apparently can be taught a form of human language that clearly involves the most abstruse relationships in the physical world.

LANGUAGE AND THE BRAIN

Despite a vast literature and recent renewed interest in this topic, it is amazing that very little is yet known about brain mechanisms of language. The facts agreed on by all are few. Language may be disturbed more dramatically by lesions to the left than to the right hemisphere, and the recent dichotic listening experiments have shown that verbal material is more likely to be processed in the left than in the right hemisphere. The frontal lobe is indispensable for motor coordination in speech, and the temporoparietal lobes are indispensable for receptive and cognitive aspects of language processing. The lateralization is not obligatory but may be blocked by early lesions to the left side. When receptive and cognitive aspects of language are disturbed, there is invariably concomitant disturbance of nonverbal cognitive processes. There are several neurolinguists who have recently claimed that human neuroanatomy may be closely related to specific language function. It is interesting to note in passing that acceptance of Premack's (e.g., 1971) and the Gardners' (e.g., 1971) demonstrations would at the same time erode the claims made by these neolocalizationist theories. (At least Geschwind (e.g., 1965) believes that the subcortical fiber connections that are unique to *Homo sapiens* are necessary for language.) Progress in neurolinguistics is basically hampered by the same difficulties discussed above. Because the nature of language is hardly any clearer than the nature of knowledge, it is difficult to know what should be correlated with anatomy and physiology.

ON THE UNITY OF LANGUAGE AND COGNITION[1]

Words and sentences concern relationships. In a verbal utterance, an assertion is made or notice is given that a certain relationship holds in the world (and the addressee is invited to check or act on it), or an inquiry is issued as to whether a given relationship holds, or a command is given that demands of the addressee that he or she bring about certain relationships—physically, for instance, by asking him or her to put *a* on *b*, or mentally by asking him or her to relate the concepts *x* and *y* in such and such a way. The language acquisition process may be thought of as a sequence of tasks in which the learner is constantly faced with such problems as "what does the word *w* stand for?" or "is it true that here is an example of *w*?" Even when the child learns such

[1] Parts of the material in this section were excerpted from Lenneberg (in press).

single words as *table, chair, uncle, father, big, green, one, up,* etc., he or she is required to ascertain whether or not certain relationships exist. The correctness of an answer to the question "is this a table?" depends on the particular relationships that must be verified–most importantly, whether the object may have a given relationship to people (real or imaginary) and their activities, such as *sitting at* or *working on.* Certain spatial relationships, such as shape, are also important, of course, as are a number of other, qualitative ones, such as stability, flatness, etc.

All natural languages have little words, such as *and, or, on, under, into, out of, after, before,* and *if.* No parent makes special efforts to teach the meaning of such words to his children. The child picks them up by observing how the language of adults relates to circumstances he is witnessing. In order to test whether a subject (child or chimpanzee) knows the meaning of these words, the following situation would have to be contrived. First it would be made sure that the subject knew the meaning of four words, *a, b, c,* and *d,* each designating some concrete object or class of objects. Then the subject would be trained to manipulate the objects according to a simple command (which might be held constant throughout the experiment). For example, all commands might be to "take" one or more objects and/or to "put" it (them) elsewhere, but each command would specify the particular relations to be brought about between the objects. In one series of commands, the objects might always be *a* and *b;* the only thing that would be varied would be the use of *and, or, on, under, into, out of, after, before.* When this had been mastered, three- or four-term commands could be given:

("take") *a* and *b* or *c* and *d;*

a or *b* and *c* or *d;*

a after *b* and *c;*

a before *b* if *c* on *d;*

etc.

In order to be sure that the subject had learned the meaning of the relational words, it would also be wise to teach him or her to watch the experimenter do a series of manipulations and then require him or her to insert between the words *a, b* that symbol which described appropriately the relationship that applied. (If the experimenter always picked up *a* and *b* together, for example, the subject would have to place the word "and" between *a, b.*)

The subject who has learned to use a word correctly has learned to deal with the world in a prescribed way (to conceptualize the world). He or she has learned to perform certain cognitive operations on potentially available data–a procedure that can properly be called a "computation." The important point here is that learning the meaning of even single words is not a passive process. Word learning requires intellectual activity, the operation of physiological processes.

The popular model for word acquisition as an "association between a visual and an auditory pattern" is misleading, because the stimulus patterns that are frequently involved are, in the first place, so variable and their common denominator is so abstract (e.g., *animal,* or *toy*), that the pattern recognition process itself can be regarded only as the end product of a set of formidable computations. Moreover, as is well known, at least half the vocabulary refers to conditions, situations, usages, and so on that are determinable only by acts of interpretation of not just one but a host of disparate stimulus patterns (e.g., the meaning of such words as *wanting* or *hitting* or *big*). The role that the speech community plays during language learning is to make it clear to the subject that there are conditions in the world to which the words *animal, wanting, hitting, big,* or *and, if, on* refer; the community may indicate that "here is a case where the word *w* applies." It is up to the learner to discover the particular computational processes that result in the "correct solution" of the task given him, e.g., "Is this an example of *animal?*"

Knowledge of word meanings entails specific intellectual, computation-like activities. If there is evidence that someone knows a word, it must be assumed that there is a specific neurophysiological (or neurochemical) process that comes into play while the subject determines whether *x* is an example of the word *w.* When words are being related to one another in utterances, there is a whole configuration of neurophysiological patterns, because the processes themselves must be related to one another–which, in turn, is also a physiological process. This can be thought of as a dynamic gestalt of processes.

One of the most characteristic aspects of language is its relative "fuzziness." There does not exist a single vocabulary item (I am excluding proper names) the meaning of which is sharply defined. This is true of the so-called "content words" as well as of the so-called "functors" (Lenneberg, 1971). Similarly with phonology: no phoneme has absolute acoustic boundaries and identification of phonological units is always based on relational computations. These facts have been discussed many times before. In recent years it is significant that the same sort of fuzziness has been discovered in the realm of syntax, where it is now clear that even the most primitive syntactic elements are not sharply delimited against each other (which means no less than that the law of the excluded middle does not hold in syntactic analysis–a serious block for successful formalization; for details, see Ross, 1974a, b, and his notion of the category squish).

It is precisely this fuzziness that makes it impossible to separate syntax from semantics, thus wedding the realm of language irrevocably to the realm of knowledge in general. This point may be illustrated by reference to one of Chomsky's famous examples:

(1) the shooting of the hunters

This phrase is ambiguous, whereas

(2) the raising of flowers
(3) the growling of lions

are not. Can this phenomenon be accounted for on purely syntactic, that is, wholly formal, grounds, as many linguists apparently thought at one time? The early explanations seemed to point in that direction; Phrase (1) may be construed as a transform derived from either of two different source sentences, namely *hunters shoot* or *NP + shoot hunters;* whereas Phrase (2) can be derived transformationally only from *NP + raise flowers* and Phrase (3) only from *lions growl; *flowers raise* and **NP + growl lions* are clearly impossible. The fact that *raise* and *growl* were intransitive verbs was considered to be accidental to the argument. What mattered was that what was called the *kernel* in those days did not contain the starred sentences (or their respective schemata) and therefore could not undergo the further operation, the transform to a phrase of the type *the + V + ing + of + the + N*. Linguistic concepts and terminology have changed since those days, but the basic problems have not. The grammar is now said to have a semantic component that interprets sentence schemata (phrase markers) and their associated words, the latter two having been generated by what is now called the "base." Details aside, the question remains whether grammatical knowledge is separable from knowledge of the world. Linguists seem to be divided on this issue, although many sound as though they are confident that grammar as a whole (syntax and semantics) can be formalized and can be treated independently from what a speaker knows about the world. Personally, I have doubts about this and believe that grammar in general, and syntax in particular, interact inextricably with general problems of knowing. Here are some of my reasons.

Take the phrases

(4) the jailing of the thieves
(5) the bullying of the thieves
(6) the stealing of the thieves
(7) the stealing of the women
(8) the stealing of the infants

In this series, Phrases (5) and (7) are clearly ambiguous, whereas Phrases (4) and (8) are hardly ambiguous (both being more strongly related to NP + V + N than to N + V); conversely, Phrase (6) is not very ambiguous, because it is more closely related to N + V than to NP + V + N. In the old conceptual frame, it may have been argued that speakers know that the kernel is not likely to contain the source sentences *thieves jail, they steal thieves,* or *infants steal.* Of course this is merely a statement on likelihood of occurrence, not on grammaticality, and yet it is this sense of likelihood, based on knowledge of the world, that influences the structural interpretation of sentences.

It may be objected that Phrases (4), (6), and (7) are structurally ambiguous even though they are semantically fairly unambiguous. The problem is far more complex. Take the sentences used to demonstrate the need for postulating deep structures:

(9) John is eager to please
(10) John is easy to please

It is clear that their grammatical structures are different, because John is the subject of Sentence (9) but the object of Sentence (10). Because neither of these sentences is structurally ambiguous in the sense that Phrases (4), (6), or (8) can be said to be, it must be assumed that there is something in the words "eager" and "easy" that determines the structural interpretation here. It is tempting to say that the difference is that "eager" can refer only to a person and "easy" only to a task. This, in turn, may seem as though "easy" is structurally different from "eager" in terms of their "feature composition." Such a formulation merely conjures up new problems. First, it implies that the words belong to different form classes. If this is accepted, Sentences (9) and (10) can no longer be said to have similar surface structure (making the surface–depth distinction less sharp). Second, it makes it awkward to account for such sentences as

(11) John is wonderful to teach

which is both semantically and structurally ambiguous [allowing a construction analagous to either Sentence (9) or (10)]. Third, it introduces *feature* as a veritable *deus ex machina:* any kind of structural or semantic problem can be "explained" by taking recourse to semantic entities the semantic status of which is left semantically undefined.

At any rate, the dividing line between syntax and semantics does not seem to be very sharp. It has not yet been proved that syntactic decisions (such as how to interpret the grammatical structure of a sentence) can, in most significant cases, be made without recourse to knowledge in general.

It may be well to take a closer look at syntactic categories. It would be a mistake to regard *sentence, noun phrase, noun, verb, adjective,* etc., as absolute constructs. Instead, these terms are the names of relations between concatenated words. Such a word as *green* is no more an adjective, a verb, a noun, or a noun phrase when it appears in isolation than it is a subject or a predicate (viz. *The Greening of America* and *Green is beautiful,* etc.; John Updike has one of his characters in *Rabbit Redux* say: "Your mother stiff-armed Janice . . . ."). The fact that a given language, such as English, has certain words that by common usage are hardly ever allowed to behave toward other words in anything but a nounlike fashion should not obscure the fact that, formally and structurally, words per se are neutral with respect to these categories. Take a string of words such as

(12) Scotch–police–sleep–like–fish

With a bit of imagination, it is possible to impute at least as many different phrase or sentence structures to this string as there are words in it. Notice how each word in the string may assume a verb function, and many of the words may also behave as adjectives, adjectival phrases, nouns, and noun phrases.

The syntactic relationship implied by designating a word as a verb or as a noun has in natural languages acquired an, alas, vague semantic significance. Nouns are more likely to imply steadiness or things, whereas verbs more often seem to be linked to action and adjectives have something to do, in many instances, with quality. However, these semantic implications are incidental and, in fact, frequently inconsistent (*motion* is a noun, *calm* can be a verb); they are irrelevant to deciding whether a given word should behave as a noun or as some other syntactic category (that is, if the nounform or the verbform should be chosen for a given meaning). Verbhood or nounhood does not seem to accrue from semantic considerations but from rather formal properties—the manner in which the word relates to other words in a string; this relationship is formally marked either by affixes in inflecting languages or by position within the string, or by a combination of the two in analytic languages. When a child develops language abilities, he or she develops a capacity to relate words to one another after a pattern; the various types of relationships, each roughly corresponding to a computation-like process, become gradually more and more differentiated. From an initial coarse commutative relationship (where it is hard to tell in a two-word phrase which of the words is the topic and which the comment), more specialized categories emerge (more distinct topic and comment, that is, predication). This differentiation history is obscured by the fact that the words the infant is using are interpreted by the adult as verbs, nouns, etc., although they may, in fact, not yet be fully functioning as these relationships function in adult language.

There appears to be a close affinity between the gradual development of semantics and that of syntax in the course of language acquisition.

When the child first begins to use single words, it is very obvious that the semantic field of each word is very large and coarse. Such words as *car, daddy, cookie* seem to apply to a fairly undifferentiated class of objects, and the criteria for what can be called a *car* and what cannot are neither sharp nor rigorously applied by the beginning speaker. For a while, he or she adds to his or her vocabulary without, apparently, bothering to refine the semantic realms of the words acquired. Indeed, these realms are still so large and undefined that a rather small total vocabulary seems to suffice to make reference to the entire world of the child. I believe that the further increase of his or her vocabulary takes place at the expense, so to speak, of the overgeneralized extent of the semantic fields of his words. The number of words is increased by letting the semantic fields of the already existing words shrink. The semantic classification system therefore becomes more and more refined, and the function of each word becomes more and more specialized.

The increase of the vocabulary and the concomitant shrinkage of the respective semantic fields is not simply a crowding of classes into a physically constant continuum, such as would take place by refining the color vocabulary more and more. This sort of "crowding" probably plays quite a small role in the development of vocabularies. What becomes refined in the course of differentiation are the essential relations that must be computed in order to ascertain whether a certain configuration in the world conforms to the conditions summarized under the word *table, mother,* etc. The relations and their respective computations become more and more exact and more and more explicit. As a result, words emerge that seem to be more purely relational than the earlier words. *Big, my, more, up* are examples. (I do not wish to call this development a "progression from the concrete to the abstract," because there is a sense in which the very first word is the most abstract of all. Because there can be no objective definition of the words *concrete* and *abstract,* little is gained by introducing these terms.)

I should like to propose that the differentiation in the field of semantics leads necessarily and "organically" to the first and most basic aspect of syntax—predication. Because of differentiation, a single word no longer refers coarsely to entire physical or social situations. No longer does the word *mommy* cover in a vague way every aspect of the familiar provider, nor the word *sock* the entire realm of sock topics. Two separate types of relations may be computed now from a single scene, one by the name of *mommy,* the other by the name of *sock.* If the child utters both words together now, it is at first difficult to decide whether the sequence is a syntactically independent concatentation or whether the sequential utterance implies a syntactic relationship, such as predication. Perhaps at the first appearance of the joining of words, such a distinction seems unimportant. It is, however, important for the understanding of further development. Suppose the child simply uttered the words as thoughts flitted through his mind, so that the concatenating principle were either association of thoughts or the sequence in which objects are seen. This process would lead to the production of word strings with no internal structure; words within one utterance would not be related to one another functionally and common syntax would therefore not emerge. If, in contrast, the joining of words is understood as the primitive beginning of syntactic relations—for instance, in the sense that two sequentially uttered words must be interpreted as an interdependent pair, one word being the modifier of the other—then a principle has emerged that is capable of generating strings of the characteristic bracketing structure Chomsky has described. Students of child language are quite unanimous in assuming that the joining of words does represent from the beginning a syntactic interrelationship, even though there is some disagreement about the exact nature of this relationship. No one seems to believe that concatenation is ever random, and the subsequent stages of language development fully justify this position. Progress in semantic differentiation leads to syntactic development, because the reduction of the semantic field of one word naturally entails the addition of specifiers and therefore leads to topic–comment constructions by means of modifying words: what about "mommy"? What about "sock"? The syntactic process of this primitive predication is actually preserved in a rather sophisticated, fully mature syntactic process, namely compounding. Gleitman and Gleitman (1970) cite an example of this process:

Volume Feeding Management Success Formula Award

(sign in a Philadelphia restaurant), which may be analyzed in terms of predicate pairs.

There is, of course, much more to syntactic structuring than the simple predicate relationship and its iterated application. Eventually, there are many different types of relationships; how large the number is no one seems to know, and there are some reasons to believe that it is not a denumerable quantity. It seems to me that this proliferation of syntactic processes is also a consequence of semantic differentiation.

This brief discussion should suffice to illustrate the affinities between language and cognition. It would be most helpful if the development of language in the apes trained by the Gardners and by Premack could be compared with the development of language in humans. Oddly enough, very little is yet known about human language development and its relation to human cognitive development. Hopeful beginnings have been made in recent years, however, particularly by Sinclair (1975), Schlesinger (1971), C. Chomsky (1969), Bloom, (1970), Bowerman (1973), Slobin (1972), and others.

There are two further points in connection with child development that have a bearing on this discussion and also on the comparison of human language with apes' capacity for language. There is wide agreement on both points.

First, the development of human thought and knowledge is independent, to some extent, of concomitant language development. Peripherally deaf children with very impoverished language are not significantly different from their hearing contemporaries up to the end of the first decade (see especially Furth, 1966, and Furth & Youniss, 1975, and references cited in these papers). Language is therefore dependent on cognition, but the reverse is apparently not true.

Second, language development in children follows—within certain limits—a predictable course of events (Brown, 1973; Lenneberg & Long, 1974). It is very curious that the apes trained by the Gardners and Premack have not been reported to follow the same learning strategies that children seem to follow obligatorily. Undoubtedly this is because of the trainers, who have chosen their own strategies rather than those occurring naturally in children. This raises the questions of why children are so remarkably constant in their language development and of whether they could acquire language presented to them in different forms. Observation on the teaching of deaf children and on their accomplishments makes me doubt this, but I hasten to admit that reliable data are lacking on this point.

REFERENCES

Atz, J. W. The application of the idea of homology to behavior. In L. R. Aronson, E. Tobach, D. S. Lehrman, & J. S. Rosenblatt, (Eds.), *Development and evolution of behavior: Essays in memory of T. C. Schneirla.* San Francisco: Freeman, 1970.

Bloom, L. M. *Language development: Form and function in emerging grammars.* Cambridge, Mass.: M.I.T. Press, 1970.

Bowerman, M. Structural relationships in children's utterances: Syntactic or semantic? In T. Moore (Ed.), *Cognitive development and the acquisition of language,* New York: Academic Press, 1973.

Brown, R. *A first language: The early stages.* Cambridge, Mass.: Harvard University Press, 1973.

Chomsky, C. *The acquisition of syntax in children from 5 to 10.* Cambridge, Mass.: MIT Press, 1969.

Chomsky, N. *Syntactic structures.* The Hague: Mouton, 1957.

Chomsky, N. *Aspects of the theory of syntax.* Cambridge, Mass.: MIT Press, 1965.

Furth, H. G. *Thinking without language: Psychological implications of deafness.* New York: Free Press, 1966.

Furth, H. G., & Youniss, J. Congenital deafness and the development of thinking. In E. Lenneberg & E. Lenneberg (Eds.), *Foundations of language development: A multidisciplinary approach.* Paris: UNESCO–New York: Academic Press, 1975.

Gardner, B. T., and Gardner, R. A. Two-way communication in an infant chimpanzee. In A. Schrier & F. Stollnitz, (Eds.), *Behavior in non-human primates,* Vol. 4. New York: Academic Press, 1971.

Geschwind, N. Disconnexion syndromes in animals and man. *Brain,* 1965, 88, 237–294, 585–644.

Gleitman, L. R., & Gleitman, H. *Phrase and paraphrase: Some innovative uses of language.* New York: W. W. Norton, 1970.

Lenneberg, E. H. On explaining language. *Science,* 1969, **164,** 635–643.

Lenneberg, E. H. Of language knowledge, apes, and brains. *Journal of Psycholinguistic Research,* 1971, **1,** 1–29.

Lenneberg, E. H. The concept of language differentiation. In E. Lenneberg & E. Lenneberg (Eds.), *Foundations of language development: A multidisciplinary approach.* Paris: UNESCO–New York: Academic Press, 1975.

Lenneberg, E. H., & Long, B. S. Language development. In J. A. Swetts & L. L. Eliot, (Eds.), *Psychology and the handicapped child,* Department of Health, Education, and Welfare, Publ. #OE 73-0500, Washington, D.C.: U.S. Government Printing Office, 1974.

Premack, D. On the assessment of language competence in the chimpanzee. In A. Schrier & F. Stollnitz (Eds.), *Behavior in non-human primates,* Vol. 4. New York: Academic Press, 1971.

Roe, A., & Simpson, G. G. (Eds.), *Behavior and evolution.* New Haven: Yale University Press, 1958.

Ross, J. R. Squishing. In E. Burtinsky (Ed.), *Festschrift for Martin Joos.* Toronto: University of Toronto Press, 1974a.

Ross, J. R. The category squish: Endstation Hauptwort. In P. Peranteau *et al.* (Eds.), *Papers from the eighth regional meeting of the Chicago Linguistic Society*. Chicago: University of Chicago Linguistics Department, 1974.

Schlesinger, I. M. Production of utterances and language acquisition. In D. I. Slobin (Ed.), *The ontogenesis of grammar.* New York and London: Academic Press, 1971.

Sinclair, H. The role of cognitive structures in language acquisition. In E. Lenneberg and E. Lenneberg (Eds.), *Foundations of language development: A multidisciplinary approach*. Paris: UNESCO–New York: Academic Press, 1975.

Slobin, D. I. Cognitive prerequisites for the development of grammar. In C. A. Ferguson & D. I. Slobin (Eds.), *Studies of child language development.* New York: Holt, Rinehart, & Winston, 1972.

Waddington, C. H. *Theoretical biology,* 4 vols. Edinburgh: Edinburgh University Press, 1967–1970.

16

Intelligence: An Invertebrate Perspective

William C. Corning
James A. Dyal
Robert Lahue

University of Waterloo

The invitation to discuss invertebrate intelligence at what was clearly a conference on vertebrate evolution and behavior presented a number of problems for us. The major difficulty was trying to ascertain what possible value invertebrate data could be to those whose primary interest lay in vertebrate intelligence mechanisms and capacities. We came to perceive our role as that of discussing the following:

1. What are the realms of inquiry—what are the evolutionary strategies for achieving intelligence that encompass both the invertebrate and vertebrate? What strategies are possibly different? Are there invertebrate models that are useful to the vertebrate researcher?
2. In order for an invertebrate model to be useful, what is a valid comparative strategy? We argue below that a comparative strategy must include multilevel (multidisciplinary) data as well as more traditional criteria for the presence or absence of learning.
3. Finally, we argue that the vertebrate researcher must be aware of invertebrate data so that more conservative conclusions are reached concerning the relationships between particular anatomical configurations and intelligence functions. Comparative data may indicate that highly dissimilar anatomies can generate very similar functions.

ASSUMPTIONS AND STRATEGIES OF INQUIRY

Intelligence and Learning

One of the problems in conceptualizing intelligence in invertebrates is that there is a tendency to impose standards that are derived mainly from vertebrate research. This is understandable because any comprehension of learning and intelligence is best elaborated at the level of higher vertebrates and the ultimate concern is to understand the intelligence of man in an evolutionary context. However, this bias can lead to some rather spurious conclusions about what intelligent mechanisms the invertebrate does or does not possess. There are many invertebrate capacities that, on the surface at least, approximate vertebrate learning and intelligence. Beritoff (1971), for example, uses what is called "image-driven" behavior tasks to compare various vertebrates. Animals are shown food in a particular locus and in later tests they are observed to demonstrate behavior that unfolds as though the food is there. The assumption is that an "image" must exist of the food in space. Spiders also show "image-driven" behavior: they have the capacity to remember the location in space of a fly (Bartels & Baltzer, 1928). Whether this represents the same neural process that a dog or a man uses is another question. It should, perhaps, continue to be remembered that the invertebrate phyla have been separated from the vertebrate for a very long time and it is not unreasonable to assume that very different mechanisms have been used to achieve an array of intelligent capacities. In other words, there have been divergent routes to similar organism–environment transactions. This convergence of behavioral adaptations has provided a variety of interesting analogs, but as we elaborate below, the assertion of a homologous function requires much more extensive analysis of the various adaptations than is currently available. Furthermore, we do not argue that a similar level of capacity has been achieved via convergence even though many of the learning categories that enthrall "rat runners" can be demonstrated in invertebrates. It may be the case that behavioral scientists have not tapped the true qualitative differences between animal groups, although there has been some significant progress in establishing methods and concepts that can uncover such differences (Bitterman, 1965).

The imposition of vertebrate biases on invertebrates predisposes thinking that what is an intelligent behavior in the vertebrate must be a useful and desirable capacity in the invertebrate. For example, the capacity to associate stimuli is probably one of the basic requirements for vertebrate intelligence but what is intelligent for the vertebrate need not represent intelligence for the invertebrate; in annelids ". . . the existence of such learning may prove to be of much more significance to the animal behaviorist than to the worm. Indeed, there is no evidence available that any worm has actually used this capacity to effect a more adequate adaptation to his environment. While it may well be that worms *can* learn associatively the demands of their normal environments seldom, if ever, actually require that they *do* so" (Dyal, 1973). (It is, of course, possible that the behavioral situation(s) in which associative learning is critical in annelid adaptations have not yet been discovered.)

For the present it can be accepted as a general assumption that invertebrate intelligence represents or is included in the set of adaptive organism–environment transactions. Distinctions between "intelligent" and "nonintelligent" behaviors can be made when more is known about the processes and mechanisms that are involved in the transaction. There are three general routes by which successful transactions can occur.

Preprogrammed morphological–physiological strategies. This strategy for existence represents evolutionary adaptations that are mostly independent of environmental input.[1] The reliance is on the genetic program to provide a sufficient array of behavioral possibilities to meet prevalent environmental contingencies. Experience may enhance or depress the use of various system programs that are available and therefore the concepts of plasticity or learning are not excluded in this category. However, the genetic readout patterns and the program specifications are determined very early and, ontogenetically, remain relatively immutable. For example, habituation (learning not to respond) may be conceptualized as a physiological process strategy that results in the behavioral modulation of "species-typical" reflexes. At the cellular level, end-product inhibition represents an example of a metabolic process that can decrease activity in biochemical pathways as a consequence of the accumulation of a particular agent.

Epigenetic strategies. These represent processes that are dependent on environmental input (perhaps at critical developmental stages) for their proper elaboration. In this case the genes provide a "dictionary" or code–reference system that can stipulate a number of species-typical mechanisms but the system relies on environmental input for normal development of appropriate neural and behavioral patterns. Inappropriate patterns may occur if the environment is drastically changed. There are many examples of the epigenetic process at all stages of the organism–environment transaction: (1) retinal cells are observed to degenerate if an animal is light deprived (Rasch, Swift, Riesen, & Chow, 1961); (2) stimulation can increase the number of dendritic spines in cortical cells (Schapiro & Vukovich, 1970); (3) deprivation of environmental interaction but maintenance of stimulus input in young cats produces adults that are not perceptually normal (Held, 1965); (4) "enriched" environments may improve cortical circulation and glial density in rats (Rosenzweig, Bennett, & Diamond, 1972); (5) brain polysome concentration is affected by stimulation (Appel, Davis, & Scott, 1967); (6) the stimulus preferences and subsequent social

[1] This assumes a normal extracellular milieu.

relationships of an animal may be biased by early experiences at critical periods (Sluckin, 1965); (7) the development of "vertical line" detectors in cat brains may be dependent on perceptual experience with vertical inputs (Hirsch, 1972). It can be argued that all structures and functions are the result of epigenesis because proper genetic readout requires particular cytoplasmic and extracellular environments. Some have tried to distinguish between physiological and psychological influences in the epigenetic process but because psychological influences must necessarily be transduced into some sort of physiological language, worrying about such distinctions does not appear to be particularly useful.

Acquired relations between system components (suprastructural). This route to intelligence assumes the existence of a structural substrate but refers to adaptive transactions that result in new arrangements of communication between various parts. The information that is stored about some experience is represented in the new relationship, need not involve any structural modification[2] and, in fact, may not be dependent from one point in time to the next on any specifically localized structures. The sum total of contributions from various portions is the information and the relative contribution from any one center may vary. This intelligence mechanism can occur regardless of whether the system has depended on (1) genetic preprogramming or (2) epigenesis during development. John's electrophysiological investigations of learning provide a good example of this relational strategy in mammals (John, 1967, 1972). John's data have led to the formulation of a "statistical configuration" theory in which ". . . the common mode of activity of large numbers of neurons in anatomically extensive systems represents information about a learned experience. Different brain regions come to share a common mode during learning" (John, 1972, p. 863). (Details of these considerations are provided below.) Processes such as these represent a means by which the integration of activities can generate new abstractions and concepts.

Learning and Learning Mechanisms

We have recently discussed some of the problems involved in attempting to define learning and have opted for a rather broad definition at this stage of our knowledge (Dyal & Corning, 1973): ". . . our domain of interest is in the characteristics of systems which exhibit a change in behavior (response, output) as a function of characteristics of the events which have preceded or followed that response (1) when it occurred on previous occasions and (2) which changes are presumed to be mediated by structural and process changes in the system." This broad definition does not exclude such things as epigenesis, immunological responses, and adaptive enzyme formation and does include the various learning

[2] By "structure" is meant such cellular modifications as increases and decreases in dendritic processes, changes in synaptic configurations that increase or decrease transmission efficacy, and changes in cell numbers.

paradigms traditionally applied in the behavioral sciences. We adopt this general definition because until we can show that the immunological response, for example, is not a cellular mechanism that accounts for system changes as a consequence of experience, then we cannot justifiably exclude it from the domain of learning research. (Mekler, 1967, for example, has proposed an immunological mechanism as a basis for memory.)

As for assumptions about learning mechanisms, evolution has apparently provided a number of divergent processes that subserve the same behavioral output and preparations can be found that can provide reasonably good data to support a variety of hypotheses about what may subserve learning. If a "cellular connectionistic" assumption is made (Kandel & Spencer, 1968), there are a number of preparations that provide supportive data. Certainly, the belief that learning involves changes in certain cells and in intercellular specifications is made secure by the impressive evidence in developmental neurobiology that portrays the nervous system as highly organized and precisely connected. Functionally, this specificity is also apparent; there is something logically satisfying about finding a particular unit in the frog that responds rather clearly to a "fly" (Lettvin, Maturana, McCullock, & Pitts, 1959) when there is concomitant behavioral evidence that frogs actually respond to flies. On the output side, electrical tickling of specific sites in mammalian brains can elicit specific behavioral sequences independent of experience (Fisher, 1956) and the ultimate in control is seen in the invertebrate sea slug, *Tritonia,* in which a single pulse popped into an identifiable cell occasionally elicits a rather complex behavioral sequence that looks remarkably like a starfish escape response (Willows, 1973). Mapping cells, anatomically and functionally, is possible and many are discovering that the invertebrates provide amenable preparations for this strategy.

Conversely, if it is assumed that learning is represented in a coherent pattern of activity instead of in a particular connection or set of connections, there are evolutionary products (perhaps only mammalian brains) that seem to operate on this principle. The earlier ablation work of Lashley (1950) provided considerable impetus for a modern movement against the cellular connectionistic doctrine and the previously mentioned electrophysiological work of John (1967, 1972) continues to produce a strong challenge. John's objections are essentially based on the following observations. (1) The results of ablation studies are inconsistent with respect to localizing learning processes; (2) Multiple-lesion studies indicate that some functions, such as learning and memory, are not eliminated after destruction of particular brain areas and that therefore the brain does not store these functions in a place.[3] (3) The results of ablations performed early in

[3]A systematic comparison of one- and two-stage lesions in rats, involving the hippocampus, amygdala, and frontal cortex, has underscored the futility of attempting to define the brain's role in learning with localized lesions (Stein, Rosen, Graziadei, Mishkin, & Brink, 1969). One-stage lesions led to an impairment of learning, whereas after a two-stage lesion there was no impairment.

development also support the above findings. (4) Observations of unit discharge patterns show that there is a great amount of variability in firing rate and that cells can respond to a variety of different modalities–the specification of information in a single neuron seems unlikely. (5) With respect to studies of "simple systems," the plasticities that are demonstrated are relatively brief and such findings may not apply to vertebrates.

As an alternative to a cellular connectionistic assumption, John proposes a "statistical configuration theory":

> The critical event in learning is envisaged as the establishment of representational systems of large numbers of neurons in different parts of the brain, whose activity has been affected in a coordinated way by the spatio-temporal characteristics of the stimuli present during a learning experience. The coherent pattern of discharge of neurons in these regions spreads to numerous other regions of the brain. Sustained transactions of activity between participating cells permit rapid interaction among all regions affected by the incoming sequence of stimuli as well as the subsequent spread. This initiates the development of a *common mode of activity*, a temporal pattern which is coherent across those various regions and specific for that stimulus complex. As this common mode of activity is sustained, certain changes are presumed to take place in the participating neuronal populations, which are thereby established as a representational system. Whether such changes are alterations of "synaptic efficiency" or not, it is assumed that the critical feature of these changes is to increase the probability of recurrence of that coherent pattern in the network. Certain types of preexisting neuronal *transactions* become more probable, but no new connections are assumed to be formed [John, 1972, pp. 853–854].[4]

Evidence supporting this theory is derived from both macro- and micro-electrode recordings in the brains of cats. Basically, the data demonstrate that during conditioning, average evoked responses in various anatomically dissimilar regions of the brain show increasing similarity in waveshape as conditioning proceeds. Electrode position is less important in obtaining this concordance than are the stimulus characteristics. Comparisons of unit discharges and multiple-unit macropotentials demonstrate a greater invariance during learning in the macro-potential readout. Further details of these studies have been recently reviewed (John, 1967, 1972). In general, the data indicate that information representing learning can best be expressed in terms of the coherent activity of widely located ensembles rather than in terms of strictly circumscribed loci.

A somewhat different view has been offered by Luria (1966) to account for the controversy between the connectionistic and the statistical theories. Luria acknowledges the difficulties posed by a strictly connectionistic view but does not agree with a Lashley (1950) "equipotentiality" explanation. The difficulty, according to Luria, is that the concept of "center" has suggested a localized

[4] From John E. R. Switchboard versus statistical theories of learning and memory. *Science,* 1972 177, 850–864. Copyright © 1972 by The American Association For The Advancement of Science.

collection of cells in one area of the brain that are responsible for a particular function. Luria reintroduces Filimonov's notion of "pluripotentialism," in which a function is assumed to involve a set of fixed connections that are multiply represented. Many parts of the system participate in executing a function but the contributions of various parts are variable. There is at any point in time, therefore, a "constellation" of centers in a dynamic system. Lashley's concept of equipotentiality assumes the *transfer* of a function from one locus to another; Luria assumes that an existing, multilevel, anatomically diffuse but specific constellation of components is reorganized to reestablish the function. After a lesion, the remainder of a constellation can be used to perform the function.

There are not sufficient data at this time to draw any firm conclusions about how the invertebrate system can learn. At present, the cellular connectionists are finding cellular specificities in invertebrates and the "neo-Lashleyites" are finding fluidities in vertebrates, differences that may represent divergent phylogenetic routes to generally similar intelligence modes. Further comparative investigations should help to resolve this issue.

Some Comparative Strategies and Problems

The reasons for emphasizing comparative tactics in learning studies have been recently discussed (Corning & Lahue, 1972). Critical for the present discussion is the special advantage strategy, in which the particular characteristics of a preparation make it especially suited for certain research aims. Typically, the preparation offers a naturally restricted neural network or one that can be restricted surgically. This "simple-system" approach implies that there is some preconception about what is to be found (Abraham, Palka, Peeke, & Willows, 1972). Therefore, the assumption that a cell can encode information makes the giant cells of many molluscan and arthropod preparations attractive. In their excellent discussion of the simple-system approach, Abraham *et al.* (1972) underscore some problems and questions that should be seriously considered in adopting this strategy. The major difficulty seems to involve the generalizability of the data. If there is a common mechanism of learning in all organisms, there must be differences in the way organisms use this mechanism. Repression and derepression of DNA coding sites may indeed be a basis of long-term changes in cellular function (Gaito, 1972) but it is hard to imagine a protozoan, flatworm, and cat using this mechanism in exactly the same way. The mechanism may subserve a change in molecular conformation at the membrane, as increase in metabolism of a synaptic transmitter, or a modified neuronal morphology, etc. It is also possible that at certain levels of analysis (synaptic pharmacology, for example), a flatworm and a cat both rely on a depleted transmitter supply to effect, say, the habituation of a particular response. However, the network architecture that is eventually responsible for behavioral habituation is completely different in flatworms and cats. Evolutionary divergence may involve the biophysical–biochemical and/or morphological realms. Finally, are the changes observed at the simple-system level generalizable to the intact animal? There is a gamble in selecting a system for special advantages, such as the brain of *Tritonia* (Willows, 1973), where there are large, identifiable cells, that are most accessible for microelectrode recording—the system may not learn. "Do we pick a behavior capable of exhibiting learning and hope the cellular analysis can follow, or do we optimize a preparation for cellular analyses and hope for the learning to hatch?" (Abraham *et al.*, 1972).

Consideration of these problems in our own invertebrate research has led us to propose that a reasonable approach is to define and explore behavioral plasticity on the basis of both multidimensional (polythetic) and multilevel criteria (Corning & Lahue, 1972; Dyal & Corning, 1973). The polythetic approach to definitional problems within biology and psychology has been outlined by Jensen (1967) following Sokal (1966). It is contrasted with the traditional "monothetic" approach, which attempts to define the "essence" of a concept by the use of a small number of critical features. The polythetic approach requires that multiple features be specified and assigns degrees of affinity between phenomena or categories based on the the degree of overlap among the sets of features rather than by congruence of small subsets of critical features. It is asserted that the advantage of the approach to taxonomy is that it is more "natural" and more "useful." Now, the utility of a particular concept depends on its ability to organize a domain of observations and to lead to novel observations. Therefore, the utility of a concept is in part to be judged by its integrative capacity. Furthermore, the integrative power of a particular concept for taxonomy is greater to the extent that it permits several levels of analysis to be related (e.g., neuroelectrical, neurochemical, and behavioral). For example, quite in the spirit of polythetic operationism, Thompson and Spencer (1966) advance nine criteria as constituting an operational definition of habituation. This definition is multidimensional but its power is limited when analysis is only at one level. Thus, it is quite conceivable that all these criteria can be met in two different preparations and yet the mechanisms subserving the response decrement may be different (see Table 1). From the perspective of the scientific enterprise as a whole it seems clear that viable theories of learning require integration of several dimensions at several levels of analysis; i.e., they must be polythetic and multilevel. In this process a generic term (habituation) may be elaborated by defining many subspecies of habituation; habituation 1, 2, 3 . . . *n* which may refer to polythetic definitions at various levels of analysis. At each level of analysis (molecular, cellular, intercellular, component relationships, etc.) different properties are likely to be observed; i.e., at the molecular level changes in DNA readout may occur; at the cellular level structural modifications may occur; at the intercellular level alterations in synaptic relationships may take place, and so on.

Finally, some general problems of the comparative approach need to be considered. Students like to think of evolution as producing organisms of increasing complexity as well as producing organisms that are attuned to particular environmental transactions. Comparative investigations that demonstrate true differences in capacities between different species and subspecies are difficult to carry out. Animals differ in their sensory and motor capacities, they respond to different motivational factors, and endogenous factors such as hormonal fluctuations are different. How can animals possibly be equated in a particular learning paradigm to determine whether they are the same or different? The approach taken by Bitterman (1965) looks to be one useful strategy. Instead of being based on a numerical measure (trials to criterion, errors, etc.) comparisons were made on the basis of functional relations. For example, it is possible to compare rats and fish in a spatial reversal task; both animals can be trained to enter the right of left side of a two-part chamber if they are reinforced for just one choice. Reversal of the reinforced side tests a particular functional capacity and fish have been found to be inferior to rats. The argument that this difference represents a difference in motivational state or in other uncontrolled factors can be answered. If hunger level is the differentiating factor, then by varying the level of food deprivation throughout the functional range of the variable a point should be found at which the fish are equal to or better than the rat. This strategy assesses the presence or absence of a logical operation in a particular species and is not unlike the approach taken by Piaget in humans (Flavell, 1963), although it is considerably more systematic.

Assessments of capacity differences that are related to complexity differences can be carried out by comparing simplified and complex versions of the same system (Corning & Lahue, 1972; Lahue & Corning, 1971a,b). Certain invertebrates have proved to be excellent preparations for this sort of comparative strategy, especially many arthropods in which ganglia are readily identifiable and surgical separation of system components is accomplished with relative ease. This strategy does not permit any comparisons to be made between different phyletic groupings but it does permit comparisons between levels of complexity or the degree of neural relationships. The difficulties encountered in trying to equate the task for different species is obviated because the receptors are all within the same system; the stimulus input, at least, can be assumed to be constant for simplified and complex versions of the same system. An application of this strategy has recently been reported in *Limulus polyphemus* using habituation as the learning paradigm (Lahue & Corning, 1971a; Corning & Lahue, 1972). Using intact and isolated nervous system preparations it was found that the acquisition of a response decrement was less dependent on the degree of neural organization than was the retention of the decrement. Retention of habituation was poorest in single, isolated ganglia or in half-ganglion preparations. Such examinations of the relationship between complexity and learning can be conducted with "simpler" forms, such as the colonial ciliates.

TABLE 1

A Comparison of Habituation Criteria and Mechanisms[a]

		Criteria[b]	Possible mechanisms
Protozoa			
Spirostomum ambiguum	1	Wawrzynczyk (1937); Kinastowski (1963a, b); Applewhite and Morowitz (1966, 1967)	Rate of habituation is independent of temperature, protein synthesis, and RNA synthesis. A passive intracellular diffusion of ions is thought to result in an *uncoupling of contractile elements.* Recovery is temperature and metabolism dependent, probably involving active ion transport (reviewed in Corning & Von Burg, 1973)
	2	Wawrzynczyk (1937); Kinastowski (1963a); Applewhite (1968)	
	3	Kinastowski (1963a)	
	4	Kinastowski (1963a)	
	5	Kinastowski (1963a)	
	8B	Applewhite, Gardner, and Lapan (1969): a more intense or a different stimulus applied after habituation training elicited a response	
		Wawrzynczyk (1937): recovery of response was obtained following a reduction in stimulus frequency	
Stentor coeruleus	1	Wood (1970a, b), Harden (1973)	Electrophysiological recordings indicate that habituation is due to a *decrement in a sensory receptor* function (Wood, 1971).
	2	Harden (1973), Wood (1970a, b)	
	3	Harden (1973), Wood (1970a, b)	
	5	Wood (1970a, b)	
	7	Wood (1972): spatial generalization to the test stimulus, which is graded with distance from the habituated site and dependent on the initial level of response decrement	

	8A	Harden (1973): stimulation by changing the water in the test bowl resulted in an enhancement of the habituated response	
	8B	Harden (1973): a different stimulus (strobe light) resulted in a response following habituation	
		Wood (1970a, b): following habituation a different or more intense stimulus elicited a response; however subsequent presentation of the original stimulus did not elicit a full response	
Coelenterata			
Hydra	1	Rushforth, Burnett, and Maynard, (1963), Rushforth (1965, 1967, 1973)	Stimuli drive a single event generating pacemaker that *inhibits a spontaneously active multiple-event generator*—successive stimuli gradually fail to excite the single-event generator
	2	Rushforth (1967)	
	3	Rushforth (1965, 1967)	
	4	Rushforth (1965, 1967)	
	5	Rushforth (1965, 1967)	
	7	Tests failed to demonstrate any generalization; Rushforth (1973)	
	8B	Rushforth *et al.* (1963): application of a different stimulus resulted in a response	

(*continued*)

TABLE 1 (*continued*)

		Criteria[b]	Possible mechanisms
Annelida			
Polychaeta			
Nereis pelagica	1	Clark (1960a, b), Evans (1969)	*Synaptic failures* occur at the sensory-giant fiber and giant-motor junctions (Horridge, 1959). Decerebration has no effect.
	2	Clark (1960a, b)	
	3	Clark (1960a)	
	4	Clark (1960a)	
	5	Clark (1960a), Evans (1969)	
	7	Clark (1960a): no generalization across modalities	
	8B	Clark (1960a): lack of cross-modal generalization implies this type of dishabituation	
Oligochaeta			
Lumbricus terrestris	1	Kuenzer (1958), Gardner (1968)	Habituation to tactile stimuli localized at *internuncial sensory-giant and giant-motor junctions* (Roberts, 1962). Decerebration lowers both the threshold to stimulation and the rate of habituation
	2	Kuenzer (1958), Gardner (1968)	
	3	Kuenzer (1958)	
	5	Kuenzer (1958)	
	7	Kuenzer (1958): spatial generalization but no cross-modal	
	8	Kuenzer (1958): lack of cross-modal generalization implies this type of habituation	
Hirudinea			
Macrobdella decora	1	Ratner (1972)	
	2	Ratner (1972)	
	3	Ratner (1972)	
	4	Ratner (1972)	

Arthropoda		
Chelicerata		
Limulus polyphemus	1 Lahue and Corning (1971a, 1973), Corning and Lahue (1972)	An *interplay of at least three functional systems* that respond to the stimulus results in either habituation, sensitization, or both dependent on experimental parameters. Change in temperature drastically affects rate of habituation
	2 Lahue and Corning (1971a, 1973), Corning and Lahue (1972)	
	3 Lahue and Corning (1971a, 1973)	
	4 Lahue and Corning (1973)	
	6 Lahue and Corning (1971a)	
	7 Lahue and Corning (unpublished data, 1973): spatial generalization	
	8A Lahue and Corning (unpublished data, 1973): following a different stimulus the response to the habituating stimulus recovers	
	8B Lahue and Corning (1971a, 1973): a different stimulus elicits a response following habituation	
	9 Lahue and Corning (unpublished data, 1973)	
Crustacea		
Procambarus clarkii	1 Bruner and Kennedy (1970)	In an isolated motor giant-muscle preparation, habituation correlates with a *decrease in excitatory junctional potential amplitude*; not caused by a change in the electrical properties of the postjunctional membrane, inhibitory buildup, or presynaptic transmitter depletion
	2 Bruner and Kennedy (1970)	
	4 Bruner and Kennedy (1970)	
	8A Bruner and Kennedy (1970): interpolation of a short high-frequency stimulus causes the response to the habituating stimulus to recover	
	8B Bruner and Kennedy (1970): an increase in stimulus frequency elicits a response following habituation	

(*continued*)

TABLE 1 (*continued*)

		Criteria[b]	Possible mechanisms
Procambarus clarkii	1	Krasne (1969), Krasne and Woodsmall (1969)	A gradual *decrease in the size of the beta component* of the EPSP between primary afferents and primary interneuron (Zucker, Kennedy, and Selverston, 1971); deemed to be *presynaptic decrease in transmitter release* and is not used by inhibition (Krasne & Roberts, 1967); insufficient to explain behavioral habituation (Krasne, 1973)
	2	Krasne (1969), Kranse and Woodsmall (1969)	
	3	Krasne (1969)	
	4	Krasne (1969)	
	5	Krasne (1969)	
	7	Krasne and Woodsmall (1969)	
	8B	Krasne (1969): a more intense stimulus following habituation elicits a response	
Insecta			
Schistocerca gregaria	1	Horn and Rowell (1968)	Thought to be a depression in excitatory synaptic function caused by *decreased transmitter release*
	2	Horn and Rowell (1968)	
	3	Horn and Rowell (1968)	
	4	Horn and Rowell (1968)	
	7	Horn and Rowell (1968): very little demonstrated	
	8A	Rowell (1971)	
	8B	Horn and Rowell (1968)	
	9	Horn and Rowell (1968), Rowell (1971)	
Mollusca			
Limnaea stagnalis	1	Holmgren and Frenk (1961), Cook (1971)	Actively incrementing, *hyperpolarizing influences* decrease synaptic efficacy (Holmgren & Frenk 1961)
	2	Holmgren and Frenk (1961), Cook (1971)	
	3	Holmgren and Frenk (1961)	
	4	Cook (1971)	
	6	Holmgren and Frenk (1961)	

	7	Cook (1971): intramodal generalization	
	8A	Cook (1971): interpolation of an extra stimulus of a different modality causes a recovery in response to the habituation stimulus	
Aplysia	1	Pinsker, Kupfermann, Castellucci, & Kandel (1969), Castellucci, Pinsker, Kupfermann, & Kandel (1969)	A decrement in EPSP amplitude caused by a change in synaptic efficacy resulting from either a *decrease in transmitter release* or a *decrease in postsynaptic sensitivity*
	2	Pinsker *et al*. (1969), Castellucci *et al.* (1969)	
	3	Carew *et al.* (1972)	
	4	Pinsker *et al.* (1969), Castellucci *et al.* (1969)	
	5	Pinsker *et al.* (1969), Castellucci *et al.* (1969)	
	8A	Pinsker *et al.* (1969), Castellucci *et al.* (1969): a different stimulus results in a response recovery to the habituating stimulus	
	9	Pinsker *et al.* (1969), Castellucci *et al.* (1969)	
Cat	2	Thompson and Spencer (1966), Wicklegren (1967a, b)	Effects of polysynaptic low-frequency depression (Thompson and Spencer) or a polysynaptic analog of PTP at inhibitory synapses (Wicklegren, 1967a, b)
	3	Thompson and Spencer (1966), Wicklegren (1967a, b)	
	4	Thompson and Spencer (1966), Wicklegren (1967a, b)	
	5	Thompson and Spencer (1966), Wicklegren (1967a, b)	
	6	Thompson and Spencer (1966), Wicklegren (1967a, b)	

(continued)

TABLE 1 (*continued*)

	Criteria[b]	Possible mechanisms
7	Thompson and Spencer (1966), Wicklegren (1967a, b)	
8A	Thompson and Spencer (1966), Wicklegren (1967a, b) A more intense stimulus resulted in an increased response following habituation.	
9	Thompson and Spencer (1966)	

*Criteria for habituation as outlined in Thompson and Spencer (1966):

1. Response decrement to repetitive stimulation (habituation);
2. Spontaneous recovery;
3. Potential of habituation;
4. Frequency effects;
5. Intensity effects;
6. Below zero habituation;
7. Stimulus generalization;
8. Dishabituation;
9. Habituation of dishabituation

Note that 8 has been subdivided. 8A refers to a facilitation of an habituated response to the test stimulus following a dishabituation stimulus. 8B refers to a simple response recovery to the dishabituation stimulus alone.

[a]This table has appeared previously in a similar format (Lahue, 1973b).

Invertebrate–Vertebrate Differences

The tendency to equate animals on the basis of performance is readily inhibited when investigation goes beyond behavior to the underlying structure and physiology. The tremendous convergence of function that is seen in evolution has already been mentioned. The neural mechanisms and system organizations of the invertebrates differ from the vertebrate in a number of ways. Those differences stressed by various writers include the following (Bullock & Horridge, 1965; Cohen, 1967, 1970; Horridge, 1968; Kennedy, Selverston, & Remler, 1969; Vowles, 1961).

Cell morphology. Invertebrate cells are commonly unipolar[5] with the cell body located at the "rind" or periphery of the ganglion; their location may remove them from any direct participation in the integrative activity of the neuron. Most of the integrative activity must be assumed to occur in the central neuropile "core," a network of dendrites, axonal endings, and synapses. In most invertebrate preparations, for example the soma is structurally and functionally isolated. The functional significance of this is not as yet clear but it may indicate a removal of genetic apparatus from any influence or feedback from neuropile activities. The cell body "rind" can be removed in the invertebrate without interfering with neuronal conduction and integrative properties (in some cases, up to several months–see Wine, 1973). Cohen (1972) suggests that this leads to more stereotyped patterns because the genome is less likely to have its readout altered as a consequence of experience. This may also account for the difficulty in finding long-term memory in the invertebrate. The molluscs represent some exception to this picture: cell bodies are invaded by electrical activity and long-term retention is observed.

Specification of cell type, structure, and location. Cells are readily "identifiable"; there appears to be remarkable consistency within some species with respect to the function and location of particular units. This property has enabled Willows (1973), for example, to map units, that is, to determine their location in the brain and to list their functional interconnections. Whether this indicates a high degree of genetic preprogramming with little developmental modulation (epigenesis) remains to be determined.

Nerve process differentiation. In the vertebrate the dendrites are thick (fleshy) and bushy (complex), interdigitate (numerous spines), and contain typical cell organelles (mitochondria, intracellular membranous structures, etc.); axonal–dendritic differentiations can be made. In the invertebrate, the processes are less differentiated, there are fewer identifiable tracts, and the processes are

[5]"Unipolar" refers to a neuron with a single emerging process. In the invertebrate, this process typically ramifies widely and distinctions between dendrites and axons are more difficult.

long and thin and ramify widely after emerging from the cell body; they do not appear to have the protusions and inclusions that characterize the vertebrate cell and do not appear to have the array of specific synaptic types seen in the vertebrate. Processes may be seen making multiple contacts with other processes. Generally, there is a lack of ordering and grouping for the processes beyond the initial sensory processing stages.

Feedback circuitry. In vertebrates, feedback at all stages of information processing and response execution is typical; in the invertebrate this is not a typical functional property. Animals are more easily "triggered" and maintain the sequence longer in the absence of any corrective environmental feedback (Limulus, for example, continues mating behavior immediately after being pulled out of the ocean as long as an egg laying female is present). However, primitive firststage feedback operations that serve to enhance novelty detection and sharpen input are known.

Glial presence. Heavy myelination with nodes of Ranvier are present in the vertebrate; glial cells that may act as biochemical adjuncts are also present. In the invertebrate, myelination is less, the nodes are virtually unknown, and according to Horridge (1968) the functional relationship between glia and nerve is obscure, although Levi-Montalcini and Chen (1969) have found that in tissue cultures the glia appear to have a trophic function in nerve growth. This lack of heavy glial wrapping in the invertebrate may permit a more readily established functional contact between cells, although as mentioned above the lack of involvement of the genome may restrict these contacts to a "short-term" status in memory. The lack of glial enclosures also permits the possibility of extra-synaptic influences between neurons.

Neuron numbers and general organization. The numbers of neurons and their general organization should be mentioned. The invertebrate possesses fewer cells and in most systems these cells are positioned in ganglia that are often separated. Insects, for example, are estimated to have 100,000 to a million neurons in their central nervous system. However, a small percentage of these can be concerned with learning functions, for 80% of the cells are involved with the processing of sensory input (Hoyle, 1970). The brain of the crayfish (*Procambarus clarkii*) has a total of 70–80 thousand neurons. The ridiculous is achieved in rotifers where a total of 200 cells are found in the brains of some species and 800–900 cells in the entire nervous system. However, this low cell number does not prevent learning (Applewhite & Morowitz, 1967).

Integration and information processing. Vowles (1961) has considered the question of whether the invertebrate is really a simpler version of the vertebrate and stresses the following: there is a simpler neural organization on the receptor side leading to a reduction of information that is available to the central nervous system; the number of motor neurons controlling each muscle fiber is small and much of the integration of excitatory and inhibitory events occurs in the muscle tissue rather than in the central nervous system; the motor neurons are involved in many different functions and complex patterns are achieved through temporal sequences instead of simultaneous spatial patterns; in the vertebrate the fine control of patterns is carried on by the brain, whereas in the insect the ganglia seem to regulate excitability but are not involved in coordination of patterns, a process that occurs at the periphery.

Whether information processing and the integration of information from various inputs can occur via what has been previously alluded to as "suprastructural" processes remains to be thoroughly explored in the invertebrate. Certainly, the generally accepted view is that there is little or no macropotential, slow-wave activity in the invertebrate system (Bullock & Horridge, 1965). However, Hoyle (1970) suggests that large spike potentials of up to 30 m V and 5 msec in duration probably represent the activity of synchronously discharging cells. In extracellular recordings of the neuropile, Hoyle also finds slow negative potentials that precede and accompany bursts of muscle potentials. These could represent summated excitatory postsynaptic potentials (EPSP's) and inhibitory postsynaptic potentials (IPSP's) that bias the thresholds of motor cells. This type of slow-potential involvement in integrative activities may represent the ultimate in invertebrate suprastructural possibilities, but the critical difference between the invertebrate and vertebrate may be that in the former the process is still anatomically locked, whereas in John's data (John, 1972) the macropotentials can be generated by dissimilar structures.

The preceding morphological and physiological differences suggest that the invertebrate and vertebrate very likely use differing substrates and mechanisms in learning. In a consideration of invertebrate characteristics, Cohen (1972) concludes that with respect to plasticity, the invertebrate's capacity to learn is limited by the lack of effect of electrochemical events on the genetic apparatus (or vice versa), and the smaller numbers of cells. In spite of these limitations, the invertebrates have been shown to do remarkably well, as we hope to indicate in the following sections.

INVERTEBRATE CAPACITIES

Examples of "Vertebrate-Like" Functions in Invertebrates

There are a number of functions that have been the subject of neurophysiological and psychological interest that are typically associated with the vertebrate central nervous system. However if the invertebrate physiological and behavioral literature is examined, there are many examples in which the invertebrates have arrived at the same function with divergent nervous systems. Some of these examples are the following.

Equipotentiality. Lashley's (1950) principles of "equipotentiality" and "mass action" derived from studies demonstrating that some complex psychological functions, such as learning, could not be localized in a part of brain. The decrement in performance produced by a lesion was found to be proportional to the amount of tissue removed, not to the locus. Furthermore, ablation-behavior data frequently suggest a recovery of function after the removal of a "locus" for a particular function. Another example of this effect in mammalian brain was provided by Adametz (1959), who ablated midbrain tegmentum in two to three stages rather than in one and did not find the comatose animal produced by a one-stage lesion (Morduzzi & Magoun, 1949). There are many other examples of an apparent transfer of a function after damage (see John, 1967, for a discussion). Whether this is a transfer or the utilization of the remainder of what Luria called a "constellation" is not known but there are many invertebrates that demonstrate a similar process. In *Blatta orientalis*, antenna cleaning is carried out by the front legs. If the front legs are broken off, the cleaning response returns 8 days later, when the roach stands on three legs and cleans its antennae with the fourth (Luco, 1971). This transfer of function may actually represent more of a Luria "constellation." There may be circuit redundancy so that the function is not actually transferred; that is, the available and relevant circuitry exists further back in the nervous system and what the animal learns is to balance itself again. The neural paths responsible for this equilibrium learning have been studied and electrophysiological facilitation in certain units after leg removal has been observed. Young roaches show the same transfer phenomenon and when the leg regenerates, the normal front leg wiping pattern returns. If the leg is then pulled off once more, the "transfer" occurs immediately. There are examples of this in spiders, notably the observations of Szlep (1952) in *Aranea diademata*. The cobweb of a spider is a complicated structure to effect and, as in the roach, the functions of amputated legs can be assumed by other legs.

Acquired responses also demonstrate the spread of information within the central nervous system. In Horridge's ingenious roach preparation (Horridge, 1962), the animal or a part of the animal (isolated ganglion preparation) could learn to keep one leg flexed in order to avoid shock. When other legs are tested after training, there is a savings in learning. The spread of acquired information is also observed in the planarian regeneration experiments (Corning, 1967): when trained animals are transected and allowed to regenerate, there is a savings in retraining in both halves. This partial retention is equal for the anterior and posterior regenerates and is observed in classical conditioning and maze discrimination as well. A transfer of habituation has been observed in *Limulus* (Corning & Von Burg, 1968), protozoans, and rotifers (Applewhite, 1968).

Dominant focus. Roy John (1967) has provided a thorough discussion of an interesting state of sustained excitability termed a "dominant focus" by Ukhtomski. Basically, a dominant focus is achieved by hyperexciting a reflex or some portion of the nervous system. The changed level of activity in an area of the system is sustained for a period of time after the original stimulation ceases and the area is able to summate input; i.e., formerly inconsequential inputs now have access to it and can interact with other inputs. As a result of this temporary excitatory state, the effects of sensory inputs and presumably the integrative possibilities are changed. For example. Rusinov (1953) applied subthreshold anodal current to an area of motor cortex and then presented the animal with a stimulus (light or sound); during polarization and for a period of time after its cessation, the endogenous stimulus would elicit the reflex appropriate to the motor area that was stimulated (for reviews see Adey, 1969; John, 1967; Morrell, 1961; Razran, 1971). Morrell (1961) established that there was some specification to this temporary connection between an external stimulus and a motor response. During application of an anodal current to the cortex, a 200-Hz tone was presented 30 times and a 500-Hz tone was presented only once. Prior to polarization, units in the motor cortex did not respond to any auditory input, but during polarization units began to respond both to 200- and to 500-Hz signals. After polarization was terminated, the units only responded to the 200-Hz tone. The "address" of sensory input would seem to depend on the area undergoing a heightened level of excitability.

The work of Franzisket (1963) on the frog (*Rana esculenta*) appeared to have demonstrated this phenomenon at a behavioral level in a "lower" vertebrate but subsequent replication attempts (Corning & Lahue, 1971) showed that the reflex enhancement and stimulus "rerouting" were nonspecific.

There are two phenomena studied in simple systems that may be relevant to the dominant focus demonstrations: posttetanic potentiation (PTP) and heterosynaptic facilitation (HSF). The two appear to be similar in that an increase in excitability alters the effect of a sensory input, but they represent different neural mechanisms. For PTP, the basic paradigm is to use a monosynaptic pathway of the cat spinal cord. A test stimulus is delivered to the dorsal root once every 2 sec. This results in a ventral root response of a particular amplitude. Then the dorsal nerve is stimulated at a high frequency for a period of time, after which the test stimulus is again applied. The response of the motor neuron pool is now enhanced and this potentiation can last for several hours, depending on the system and the particular stimulus parameters. Although PTP was initially demonstrated in mammalian spinal cord preparations, similar demonstrations were reported in the invertebrate (Bruner & Kennedy, 1970).

The phenomenon of heterosynaptic facilitation (HSF) has been extensively explored in a giant cell of *Aplysia* (Kandel & Tauc, 1965). The procedure is to first stimulate one pathway and to record the EPSP in the cell. Stimulation of another pathway after this initial "priming" stimulation was observed to increase the EPSP produced by the test stimulus by as much as 700% and for as long as 42 hr (Haigler & Von Baumgarten, 1972).

The consensus is that both PTP and HSF involve alterations in presynaptic events but that the two processes show many dissimilarities with respect to mechanisms. PTP is shorter lasting, sensitive to temperature, unaffected by Li^+

and Na^+, and involves a single pathway (potentiation of one pathway does not "transfer" to others). In HSF, the enhancement is longer, it is unaffected by cooling and by Li^+ and Na^+, and it involves multiple pathways. One current speculation is that the priming stimulus increases the activity in a third neuron that terminates on the presynaptic terminals of the test pathway.

Both PTP and HSF provide mechanisms for short-term "memory." Interpretation concerning the altered stimulus consequence following excitation should be based on knowledge of the circuitry (i.e., single or multiple pathways), the synaptic physiology, and the actual input pathways that are being activated by the testing and priming stimuli. Because PTP is an increase of synaptic transmission in a single channel, it may represent the basis of nonassociative behavioral changes (sensitization and pseudoconditioning). HSF and the dominant focus phenomenon may be closer to being true associative conditioning mechanisms because the response potentiation is produced by pairing stimuli in two different channels. Kristan (1971) has demonstrated in *Aplysia* that pairing the test stimulus with the priming stimulus produces changes that are different from those elicted by the test or priming stimulus alone. To demonstrate a closer parallel with Morrell's findings, a previously unpaired stimulus should be used to determine the specificity of the association.

Incrementing and decrementing processes.

"Learning not to respond" (habituation) is frequently referred to as a simpler form of plasticity but recent physiological and behavioral findings suggest it is quite complex. The habituation criteria listed by Thompson and Spencer (1966) provide a polythetic basis for comparing species. Critical among these criteria is the requirement that fatigue be ruled out and the usual method for demonstrating this is the use of a different "dishabituating" stimulus. Ideally, when the organism is showing a response decrement to one stimulus as a result of repetition, a different, previously unapplied stimulus causes a return of the response level to initial values. This demonstration rules out fatigue and also shows that the response decrement is specific to the stimulus characteristics. For example, it is possible to habituate a visual evoked response in cats to a ten per second flicker; i.e., the amplitude of the evoked potential decreases with the ten per second repetitive input. Alteration of stimulus frequency to six per second elicits a full amplitude evoked response again (John & Killam, 1960). The dishabituation that occurs when the hifi turntable drags momentarily or when lights dim briefly is familiar to all. Pumphrey and Rawdon-Smith (1939) have provided a most unusual example of the specificity of habituation that is possible in an arthropod. Repeated stimulation of the cercal nerve of the roach at a particular frequency resulted in the diminution of the postsynaptic response of a monosynaptic arc; the presynaptic input remained constant. The interpolation of one extra pulse in the stimulus train of 40 pulses per second resulted in a sudden increase in the postsynaptic response. The roach may also be able to appreciate turntable performance.

Habituation appears to involve more than one process. Groves and Thompson (1970) have provided an interesting review of the literature which suggests that two independent and opposing processes are operative during habituation. One process, habituation, involves a decrement in activity of the S–R pathway, whereas the opposing process, central state, sensitizes the response. The two processes interact to yield a particular response level. An increase in central state can prevent habituation from taking place; increased stimulus intensity or frequency are ways to accomplish this. Groves and Thompson cite a number of vertebrate preparations in which these two processes are observed. We have obtained good evidence for the two processes of habituation in an arthropod and, in fact, have complicated the model somewhat (Lahue & Corning, 1973). Dorsal nerve efferents in *Limulus* were activated by several different tactile stimulation frequencies. With the exception of the slowest (one per 60 sec), habituation occurred at all frequencies. The central state sensitization process was frequency dependent, being maximal at stimulation rates of two per second and one per second.

Ganglion divisions uncovered a possible third process. Previous research had demonstrated the existence of critical intraganglionic contralateral influences (Von Burg & Corning, 1970). These influences could be disrupted by a ganglion "split." The effect of this disruption on the habituation of dorsal nerve efferents was examined at a stimulation frequency that elicited clear incremental and decremental phases. In one group, the segment that first processes the tactile input was divided; in another group the segment giving rise to the dorsal nerve efferent was divided (output split); in a third, both input and output ganglia were divided; the fourth group included intact abdominal ganglia preparations. The results showed that habituation was best and central state sensitization least when an output ganglion was split. The most sensitization and poorest initial habituation was seen with an input split. The intact group and the group with both ganglia split were the same and showed response changes that lay between the other groups. The effect of an output split was assumed to disrupt the number of influences (central state) on the final common path, whereas an input split appeared to interfere with a third hypothetical process, a sensory modulation function. Dishabituation tests with a different stimulus (a water drop) showed a much higher response in preparations with an input split, providing further evidence for the existence of a sensory modulation function. Repetitive input, then, excites an S–R pathway, it adds to and may enhance central state influence on the efferent, and it increases the operation of sensory modulation processes. If central state influences are not sufficiently large, habituation occurs. The complexity of habituation, then, appears to be as great for the invertebrate as for the vertebrate.

Memory transfer. The invertebrate is not free of the more controversial aspects of brain research, in fact, in the case of "memory transfer," it was an invertebrate preparation that created interest in the phenomenon. Several years ago, McConnell (1962) discovered that there was an apparent facilitation of learning when a naive planarian cannibalized a trained animal. This transfer was later found to be effected by cell fractions (RNA) as well (Zelman, Kabat, Jacobson, & McConnell, 1963). Jacobson's group has presented data strongly indicating that the transfer of training effect is specific; naive animals receiving RNA fractions from groups receiving paired light (CS) and shock (US) demonstrate a higher response to light when tested than do animals receiving RNA fractions from randomly stimulated or unstimulated controls (Jacobson, 1971). Actually, the transfer effect was first accomplished with the colonial ciliate, *Carchesium* (Corning, 1971b; Plavilstchikov, 1928). Light and vibration pairings were observed to produce what was a conditioned response in the colony in 100–200 pairings. A portion of the conditioned colony was grafted to a naive colony and left for 7–22 trials. The graft was then removed and the host tested. It was found that there had been a transmission of excitability.

Research with such vertebrates as rats and fish has demonstrated that some sort of transfer effect is possible in more complex systems. Transfer effects have been obtained with whole brain injection, RNA fractions, and protein fractions, with the routes of injection being intraperitoneal or intracisternal (for a review, see Fjerdingstad, 1971). The failure by some to demonstrate the effect or to show the specificity of the effect (that is, whether it is a specific transfer of acquired information or a more general excitatory effect) has led to considerable skepticism, but a recent analysis of the literature indicates that the transfer effect is real although the interpretation of the effect is unsettled (Dyal, 1971). It would be easy to conclude that, because transfer effects are obtained with both the planarian and the rat, the same mechanism is involved but there are few data that clearly show what the mechanisms or the critical carriers for the transfer effect are. In fact, data that proves the injected fractions get into the brain cells of the hosts are hard to find.

Unit conditioning. One of the pioneers of unit conditioning, Olds (1963), has described a procedure by which the spontaneous firing rate of a neuron may be driven up or down depending on contingent reinforcing stimuli delivered to "positive" or "negative" reinforcing areas in the hypothalamus of cats. The operant conditioning of unit activity can also be demonstrated in the invertebrate (Hoyle, 1965). Certain muscles of a leg in the locust are controlled by a single excitatory axon. The activity of muscle units therefore reflects the output of a single neuron. Because the particular muscle (metathoracic anterior adductor) caused the leg flexion, it was assumed that increased muscle discharge frequency indicated leg flexion and a decrease indicated leg extension. Horridge, in 1962, had previously demonstrated that the leg position of a locust could be operantly conditioned. A circuit was devised that would deliver a shock to the leg whenever the muscle discharge fell below a certain rate, a paradigm similar to that used by Horridge. With these procedures it was possible to alter the rate of discharge in the muscle over a time course that was similar to that obtained during behavioral training.

Image-driven behavior. The use of "image-driven" behavior to differentiate between species has been reviewed by Beritoff (1971). After animals are shown food in a particular location subsequent behavior indicates that the animal acts as though the food were there. As mentioned in the introduction, analogous demonstrations have been reported in spiders (see Lahue, 1973a, for a recent review).

When presented with several prey at once, orb-weaving spiders return the prey one at a time to a retreat where they are bound while the remaining prey are retrieved. Observation of such behavior immediately creates suspicion that the spider remembers not only the presence on the web of an unfinished meal but also its location. Spiders continue to search for prey that have been removed (or escape) from the web. However, as the removal of the fly left a hole in the web that may have been a cue for the spider, Baltzer (1923; also Peters, 1932) carefully removed the fly without disturbing the web or the spider. Nevertheless, after being unable to immediately relocate the prey, spiders searched for as long as 47 min, repeatedly returning to the exact site where the prey was initially caught.

Working with the funnel-web building spider *Agelena labyrinthica,* Bartels (1929; Bartels & Baltzer, 1928) also demonstrated that the return to a prey site is not necessarily dependent on the spider's actually sensing the presence of the prey. In these experiments the spider repeatedly ran directly to the test site as many as 19 times in over $3^1/_2$ hr.

Unit specifications. Such studies as those of Lettvin *et al.* (1959) or Hubel and Wiesel (1963) demonstrating the existence of units in frog and cat nervous systems that respond to complex environmental configurations have stimulated considerable research in the identification and exploration of single-cell functions. Certainly, the ability of a unit to respond maximally to a small jerky black dot moving into the frog's "striking" range was a startling finding. Such units are also common in the invertebrate. To cite two examples, there are interneurons in the optic peduncle of the rock lobster that respond to various classes of movement, e.g., fast, medium, and jittery movement, and to the movement of light (Wiersma & Yanagisawa, 1971). There are CNS units in crickets that respond to moving targets but not to forced movement of the eye (Palka, 1969).

"Command units," or cells that seem to control complex sequences of reflexes or specific states, are found in invertebrates. In certain cases, single-cell stimulation will occasionally trigger a swimming escape response, even when the stimulation only elicits a single impulse or short burst in the cell (Willows, 1969). Analogous effects are seen in mammalian brains, where sleep, rage, maternal patterns, postural movements, and other behavioral sequences can be triggered by the stimulation of particular sites (Hess, 1957; Fisher, 1956).

Biochemical correlates. The finer biochemical correlates of learning in the vertebrate brain have been the subject of intensive and frequently controversial research in recent years. Both RNA and protein synthesis seem likely to alter with learning and one popular strategy has employed specific inhibitors of such synthesis to block learning. Both short- and long-term memory in fishes can be influenced with such inhibitors as puromycin and cycloheximide (Agranoff, 1967). A related tactic is to analyze brain macromolecules following learning in an attempt to identify specific molecules that may be synthesized as a result of the experience; Hyden's (1967) work on rats is illustrative.

Using the cockroach leg lifting preparation described above, workers in Kerkut's lab (Kerkut, Oliver, Rick, & Walker, 1970) have demonstrated the manner in which the two biochemical strategies complement one another in invertebrate studies. At first, metabolic inhibitors (cold temperature) and RNA and protein synthesis inhibitors (actinomycin D, Congo red, cycloheximide, etc.) were all shown to retard learning. Other experiments showed that the incorporation of labeled uridine into RNA was greatly enhanced in experimental animals. The inhibitors that slowed learning also inhibited the incorporation of label. Similar effects were demonstrated in the incorporation of labeled leucine into protein. Radioautography and sequential scintillation counting of thin sections demonstrated that the uptake of labeled uridine was greatest in that portion of the ganglion containing the interneuronal and motoneuronal connections most immediately involved in flexion of the leg being trained. Drugs known to decrease cholinesterase (ChE) activity facilitated learning. ChE levels fell during learning but gradually recovered over 2–3 days (in nondrugged animals), during which period the learned response decayed substantially. Woodson, Schlapfer, and Barondes (1972) have failed to replicate the ChE findings of Kerkut. However, experiments with the *Limulus* habitation preparation (see page 237) also indicate that a decreased ChE activity correlates with learning (Lahue, Thompson, & Corning, unpublished data, 1973).

Analysis of labeled uridine in single neurons in molluscs has given equivocal results (Wilson & Berry, 1972; Bocharova, Borovyagin, Dyakonova, Warton, & Verprintsev, 1972). However, the application of biochemical techniques to invertebrate preparations continues to offer great advantage in understanding neuronal chemistry and the effects of learning. In all likelihood more careful control of experimental conditions and standardization of analytical procedures can eliminate many of the present ambiguities.

The foregoing represents just a few of the many examples of where invertebrates and vertebrates share some similarities in processes that may be relevant to learning and intelligence. The following section provides an overview of some of the specific learning capacities that are demonstrable in invertebrate groups. It can be seen that with respect to mammalian-derived categories of learning, many invertebrates do remarkably well.

Formal Learning Demonstrations in Invertebrate Phyla[6]

Protozoans. It is clear that learning demonstrations with protozoans are far from convincing with the exception of habituation. In all associative conditioning paradigms, two general problems have occurred: first, there have been failures to replicate reported associative conditioning demonstrations; second, the "acquired" response modifications can be attributed to uncontrolled variables rather than to learning (Corning & Von Burg, 1973).

The "model" controversy is the Gelber-Jensen exchange that occurred a few years ago. Gelber (1952) had reported what looked like an approach conditioning demonstration in *Paramecium aurelia*. Essentially, her technique was to present a culture of paramecia with a wire coated with food reinforcement for a certain number of trials. The number of animals that approached or attached to the wire was counted. In naive cultures a bare wire elicited few attachments but when the wire was coated with food (the bacterium *Aerobacter aerogenes*) and exposed to the culture for 40 trials, it was found that there was a progressive increase in the number of animals attracted to it. In a final test, a bare wire elicited a significantly greater number of attachments when compared to attachments observed in naive cultures and when compared to a control culture that had been exposed to the wire without food.

The interpretation of these findings was disputed by Jensen (1957a, b), who felt that "learning" had not been demonstrated conclusively and that other interpretations were equally likely. Jensen's major points were that the repeated dipping of a wire coated with bacteria would produce a bacteria-rich area, that *Paramecia* would migrate to and remain within this area, and that the presence of bacteria would increase attachment responses (positive thigmotropism) in *Paramecia*. Therefore, the creation of a food-rich zone accounted for the increased number of animals at the point where the wire was dipped rather than an "association" between wire and food. If these interpretations are correct then stirring the culture fluid prior to the final test with a sterile wire should provide a critical test. This procedure was carried out by Katz and Deterline (1958), who found that the approach response to the wire in "trained" cultures disappeared. Gelber's acquisition curves were replicated by Katz and Deterline; in fact, there was a better experimental–control group differentiation than that obtained by Gelber. The reinforcement procedures produced a steady increase in the number of animals approaching the wire. What is damaging to Gelber's position are the data obtained from the counts made after stirring. When the "trained" culture

[6] Parts of this section have been published elsewhere (Corning & Von Burg, 1973).

was stirred and then retested, the number of attachments dropped markedly and was no different from the attachment frequency observed in controls. Furthermore, it was observed that the lowering of bacteria into the culture only at trial 40 increased the number of attachments during the final count, even though this group had never experienced "training" trials of wire–food associations.

Although the Katz and Deterline data cast doubts on the learning interpretations of Gelber, they were not entirely conclusive. In a summary of her research, Gelber (1965) reports that mixing of her cultures had been carried out, and that when trained cultures were washed and placed in a clean medium the animals still demonstrated retention.

A longer debate has revolved around the possibility of conditioned avoidance learning in various protozoans. Since the early part of the century, when the issue was over where "consciousness" appeared in evolution, there have been surges of controversies concerned with avoidance learning demonstrations. Among the earlier studies, those of Bramstedt (1935) and Grabowski (1939) adequately characterize the problems. In Bramstedt's experiments, the subjects (*P. caudatum* and *Stylonychia mytilus*) were placed in a water drop placed on the dividing line of a two-compartment chamber one half of which was heated to 42°C while the other half was held at 15°C. In addition, the warm half was exposed to a light while the cold portion was kept dark. At first subjects would spend an equal amount of time in both halves but by the end of the first hour they would be on the dark side for most of the time. Following the 90-min exposure to temperature gradient, the temperature of the two halves was equalized (both halves at 15°C) and the subjects' preferences then measured. The light avoidance was found to persist.

Grabowski (1939) attempted a careful replication of the "learning" demonstrations of Bramstedt. Using Bramstedt's basic stimulus conditions (one side of the well was lighted and heated; the other side was dark and cold) Grabowski observed similar behavior—during test sessions the organisms spent less time in the lighted half even though the temperature throughout the well was equilibrated and cool. However, he noted that the animals appeared to be reacting to a different kind of border than the light–dark one. The subjects were displaying a reaction after they passed the light–dark division and Grabowski concluded that the avoidance-producing stimulus could not be light. It was suggested that the paramecia were not reacting to light but to a chemical gradient that had been set up by the stimulus conditions and that persisted for a time during testing, for example, an oxygen deficiency or CO_2 built up. A further test of the chemical gradient hypothesis consisted of placing naive subjects into a well that had contained previously trained animals. These subjects behaved similarly to the trained ones.

Recently, Bergstrom (1968) has used the ciliate, *Tetrahymena pyriformis*, in attempts to demonstrate avoidance learning with light and shock pairings but replication attempts have failed to confirm his findings (Applewhite, Gardner, Foley, & Clendenin, 1971).

Simple escape learning demonstrations have also been controversial. French's (1940) earlier experiments have been generally accepted as indicating that paramecia can learn to escape a tube with increasing efficiency. In more carefully controlled experiments, Hanzel and Rucker (1971) have repeated these findings. However, Applewhite and Gardner (1973) have shown that this increase in escape speed can be eliminated if the capillary tubes are cleaned prior to use. Controls indicate that it is the capillary tube, if anything, that demonstrates some sort of "acquisition." Using the same tube, three different protozoans were given four trials each, in succession. A decreased escape time was observed in the subjects run last in the tube. Other controls indicated that if the subjects were first adapted to being picked up into the tube, "learning" did not occur. This particular debate is not settled yet, but in view of the past history of such learning demonstrations the issue is not likely to be unequivocally settled.

The indisputable learning phenomenon observed in protozoans is habituation. There is some controversy over whether all the criteria can be met (Eisenstein & Peretz, 1973), but no one seems to question the ability of a variety of protozoans to show habituation. The studies by Harden (1973) represent an excellent example of this capacity in *Stentor*. These experiments demonstrated that repetitive mechanical stimulation resulted in a decrement of contractability and that there was no retention of the habituation over a 24-hr period. However, retention is possible after an hour. Various controls ruled out the possibility of the accumulation of any substance in the medium that would account for the behavioral change. A dishabituation effect was also obtained—the changing of the medium transiently increased responsiveness to the mechanical stimulus. Fatigue was also ruled out by the observation that a light stimulus could elicit a full contraction at the completion of the habituation run with mechanical stimulation.

The potential of protozoan preparations in researching the multilevel correlates of a plasticity is being realized with the development of procedures to record electrical activity during habituation and with the application of biochemical techniques to explore ionic and macromolecular substrates of the behavioral change (Applewhite, 1972; Wood, 1970b). Further details of these studies are available elsewhere (Corning & Von Burg, 1973).

Coelenterates. The change from unicellular to multicellular existence has not produced any dramatic increase in learning capacity, although there have been some suggestions that crude associative conditioning is possible. In a recent review of the literature, Rushforth (1973) essentially confirms that evidence for habituation is strong and that demonstrations of associative conditioning are as yet incomplete. Much of the earlier evidence on acquired food responses in coelenterates is weakened by the lack of controls. Rushforth points to the as yet unreplicated work of Zubkov and Polikarpov (1951) as a possible demonstration of position learning in *Hydra*. In these experiments, it was shown that when *Hydra* were attached to a substrate they would show extension movements in various directions. If the substrate was moved close to the surface, the direction of extension could be restricted—the animals avoided extending toward the surface. After the animals had been left there for an hour, they were lowered again and for 3–4 hr they avoided extensions in the surface direction.

Ross (1965) has summarized various attempts at associative training with the sessile coelenterates. Ross presented *Stomphia coccina* with pairings of a starfish extract (CS) followed 2 sec later by strong mechanical stimulation (US) that caused them to contract. Before training the extract elicited swimming but after training, it elicited contraction or retraction for a few trials, followed by a return of the swimming response. Attempts to establish associative learning in the sea anemone were not quite as successful.

The extensive research by Rushforth (1967, 1973) on habituation in *Hydra* is providing some interesting behavioral and physiological data for this type of plasticity. Some species of *Hydra* demonstrate habituation to repetitive mechanical stimulation. There is retention of the decrement for perhaps 4 hr. Electrophysiological recordings in *Hydra* have demonstrated the probable existence of two pacemakers: a spontaneous contration pacemaker (CP) and rhythmic potential pacemaker (RP). Electrical potentials are also recorded in the base of the tentacles (TCP). The CP and RP pacemakers activate antagonistic systems; the former shorten the column and the latter cause elongation. Mechanical stimulation elicits CP's, which for a time replace those spontaneously occurring. These stimulation-induced CP's gradually disappear (habituate) and the spontaneous ones return. The RP is unaffected by mechanical stimulation. Light stimulation activates the RP system and inhibits the CP and TCP system. Habituation of activity in the TCP system also occurs. The interactions between various endogenous systems and exogenous perturbations is developing into a most complicated picture for a nervous system that was thought to be "primitive."

Planarians. Flatworms perhaps suffer from too much publicity. At one time, the debates over whether they could learn or not occupied considerable journal space and many researchers opted for less controversial preparations. In a recent consideration of these debates, the evidence clearly indicates that planarians, which represent a significant step in neurological advancement, are capable of learning in most paradigms (Corning & Riccio, 1970). The appearance of a true centralized nervous system, complete with head ganglia and synapses, is a significant departure from the nerve net of coelenterates. For the present purposes, examples are provided of learning demonstrations in classical discrimination conditioning and in two instrumental paradigms—maze learning and operant conditioning. More thorough reviews abound (see McConnell, 1966; Corning & Ratner, 1967; Corning, 1971a; Corning & Riccio, 1970; Corning & Kelly, 1973).

In an experiment that should have dispelled all controversy over the ability of planarians to be classically conditioned, Jacobson (1967) dealt specifically with the problems of pseudoconditioning and sensitization as alternative explanations for the earlier findings that indicated associations could be formed between light (CS) and shock (US). Comparisons of paired and unpaired light and shock treatments demonstrated a differential response change: the paired treatments increased the percent contractions to the light to a level of about 60% after 125 trials, whereas the unpaired presentation resulted in a 8–10% contraction rate to light. Backward conditioning (shock followed by light) and simultaneous presentation of the CS and US failed to generate high light responsivity. To effectively rule out sensitization, differential conditioning procedures were successfully applied.

Operant training paradigms have also been successful with planarians (Best, 1965; Crawford & Skeen, 1967; Lee, 1963). Basically, the procedure has been to require the flatworm to move through a small tunnel, a response that breaks a photoelectric beam and turns off an overhead light of bright intensity for 15 min. Rate of tunnel entry has been found to increase over time. Alternative explanations have been effectively rebutted (Best, 1965; Corning & Riccio, 1970).

Maze learning has also been a successful learning situation for planarians. In the standard T maze, planarians have been able to negotiate discriminations between right and left alleys, rough and smooth surfaces, and black and white alleys. One of the oldest demonstrations of maze learning remains one of the most convincing. Van Oye reported in 1920 that planarians were capable of learning to locate food on the end of a wire. The food (a piece of meat) was placed on the end of a wire that at first was positioned where the animals had a high probability of finding it. It was gradually lowered as training proceeded and, in final tests, planarians that had this food-location training were found in greater numbers on a bare wire that had been inserted into the container. Presumably, an association was established between the wire and the food, and the animals learned to approach the surface, crawl across it to the wire, and then move downward to where reinforcement had been presented. P. H. Wells (1967) has replicated these studies and has also reported that the wire-seeking behavior can be brought under stimulus control. Approximately 25% of the subjects can learn to approach the end of the wire only when an overhead light is on.

These examples indicate a marked shift in capacity when compared to the protozoan and coelenterate learning literature. In spite of the general consistency of the learning demonstrations and the fact that the planarian is one of the first preparations to be used in researching memory mechanisms (Corning & John, 1961; McConnell, Jacobson, & Kimble, 1959), interest in the preparation has unfortunately subsided.

Annelids. During the past century, research on the three classes of annelids (Oligochaeta, earthworms; Polychaeta, marine worms; and Hirudinea, leeches) has revealed a surprisingly rich array of capacities for behavioral modification. More detailed and extensive discussion of this research may be found in Dyal (1973). For the present purposes it suffices to illustrate some of the successfully demonstrated capacities with a brief overview of the conclusions that can be derived from the research in the area.

Extensive research by Clark (1960a,b) and by Evans (1969) has repeatedly demonstrated habituation of the rapid withdrawal response of nereid polychaetes to a variety of stimuli. They have shown that the habituation rate varies with the type of stimulus, the response system being habituated, the interstimulus interval, and the species being tested. Clark has shown that the habituation process generated by two qualitatively different stimuli are sometimes independent and sometimes interact, depending on both the stimulus similarity and the order in which the stimuli are presented. It is not uncommon to find similar specificity–generalization effects in mammalian preparations (Thompson & Spencer, 1966). Similar habituation demonstrations and parametric explorations have been reported for other errant polychaetes (Dyal & Hetherington, 1968), sedentary polychaetes (Krasne, 1965), earthworms (Kuenzer, 1958), and leeches (Gee, 1913).

Examples of aversive inhibitory conditioning (punishment)[7] are available in the experimental literature on both oligochaetes and errant polychaetes. Smith and Dinkes (1971) have recently reported such inhibitory conditioning of a learned response in earthworms. The worms were given 20 trials, one a day, in which they learned to traverse a runway tube that exited into their home container of moist earth. In the second phase a 25% sodium chloride solution was substituted for the moist earth and the worms were given ten more trials. During this aversive stimulation phase the worms' running times significantly increased, suggesting a conditioned response inhibition.

Ratner and Miller (1959) provided the first demonstration of classical conditioning in earthworms and included many of the appropriate control procedures. The UCS was the onset of a bright light, which elicited a slight rearing and a withdrawal of the anterior portion of the body; the CS was a mild vibration. Any withdrawal response that occurred during the CS–UCS interval was counted as a UCR. The classical conditioning group significantly increased the number of CR's as compared to sensitization pseudoconditioning and base rate control groups. Replications of these effects were reported by Ratner and Miller (1959) and by Wyers, Peeke, and Herz (1964).

Temporal parameters for most effective conditioning (interstimulus interval and intertrial interval) seem to be well within the range of values obtained with mammals. In addition, such associated phenomena as extinction, the partial

[7] Aversive inhibitory conditioning (punishment) has occurred whenever an animal learns to inhibit making an innate or learned response as a consequence of that response having been punished.

reinforcement extinction effect, and backward conditioning have been demonstrated in earthworms.

The most extensive experiments on instrumental conditioning were reported by Datta (1962). She trained earthworms in a T maze with shock as the aversive stimulus in one arm and the home container as the positive stimulus in the other arm. These experiments and many similar ones by other investigators yield the strong conclusion that earthworms are capable of discrimination learning but that the primary effect is within a daily training session; retention of the specific turning response over 24 hr is minimal. Active avoidance conditioning has also been demonstrated in earthworms (Ray, 1968) as a differential instrumental conditioning and extinction, but attempts to demonstrate higher forms of abstraction involved in discrimination reversal or latent learning have not been successful (Arbit, 1961; Datta, 1962). Only two successful demonstrations of instrumental learning in polychaetes have been reported (Evans, 1963; Flint, 1965) and research on instrumental conditioning in *Hirudinea* has been notable by its absence.

Investigations of the role of neurophysiological processes in behavior modification in annelids has been limited primarily to the use of ganglionectomy. From these studies it appears that the supraesophageal ganglion is probably not necessary for any of the modifications that have been discussed. In contrast to other invertebrate preparations, electrophysiological and psychopharmacological investigations of learning and retention in annelids have been few or totally absent. Future research is most certain to involve the use of annelid preparations for the exploration of the effects of protein synthesis inhibitors and facilitators. Furthermore, the special ability of some annelids to regenerate their brains makes them attractive subjects for brain transplants and collateral attempts at interanimal transfer of acquired information.

At this point we would like to reiterate our skepticism regarding the significance of these learning capacities for the biological adaptation of annelids. Although worms possess the capacity for some higher forms of learning, it may well be that this capacity is seldom of ever excercised in response to the demands of their normal environment. The capacity may be best regarded as an "epiphenomenal" consequence of a general body plan, nervous system, and ecological dynamics that place a premium on adjustment via habituation and sensitization processes. Furthermore the apparently limited capacity of annelids to retain higher forms of associated learning may in itself be quite adaptive. This point has been previously expressed by Clark (1965).

Arthropods: the Chelicerates. Among the chelicerates, *Limulus polyphemus* (class Xiphosura) represents what should be a thoroughly studied animal with respect to plasticity capacities and mechanisms. Explorations of the functional organization of the central nervous system have uncovered properties that suggest the *Limulus* may be a useful model for understanding more advanced systems. For example, there appears to be a functional redundancy throughout the CNS with respect to the relationship between cardioinhibitors and cardio-accelerators and, more importantly, these central regulatory functions are sensitive both to endogenous, rhythmic influences and to exogenous, sensory influences (Lahue, 1973a, b).

Attempts to demonstrate plasticity in various response systems of *Limulus* have so far been limited to habituation, classical conditioning, and simple instrumental learning. Smith and Baker (1960) attempted to establish an association between light and shock. The shock elicited a telson movement. Although they reported success with this particular pairing, others have not been successful. In general, attempts to demonstrate complex types of learning in *Limulus* have not been overly convincing (Corning & Von Burg, 1968; Lahue, 1973a; Makous, 1969; Wasserman & Patten, 1969). Habituation, however, is easily demonstrated at both the behavioral (Corning, 1966; Corning & Von Burg, 1968; Lahue & Corning, unpublished data, 1973) and the electrophysiological level (Corning & Lahue, 1972; Lahue, 1973a; Lahue & Corning, 1971a, 1973).

In spiders, habituation is also a readily obtainable response modification. Evidence for associative learning is also quite strong, especially with respect to food acquisition. Bays (1962), for example, was able to train *Araneus Diadematus* to associate the taste of dead flies dipped in sugar or quinine with a specific sound frequency (262 or 253 Hz, respectively). In the case of the negative stimulus (quinine), the fly was discarded by the fifteenth trail without biting. Changes in feeding habits can also be observed in laboratory-housed subjects (see Lahue, 1973a, for a review).

Arthropods: The insects. Studies of learning in insects have been largely confined to representations of five of 28 orders. These include bees, ants, and wasps (Hymenoptera), blowflies and fruit flies (Diptera), grain beetles (Coleoptera), cockroaches (Dictyoptera), and locusts (Orthoptera). Insects, as do many other invertebrate phyla, offer to the neurobehaviorist the advantage of a greatly simplified nervous system (e.g., the "typical insect" nervous system has only about 10^5–10^6 neurons, of which perhaps only 20% are responsible for functions other than analysis of sensory input). In addition, as noted in Alloway's recent review "the segmental organization of much of the insect nervous system offers the prospect of being able to localize at least some forms of learning in a single segmental ganglion (in addition) insect metamorphosis offers another possible approach to the localization of cellular changes associated with learning and memory storage" (Alloway, 1973).

Several basic learning phenomena have been demonstrated in insects, including habituation (both at behavioral and neural levels) classical conditioning, and several varieties of instrumental conditioning. In addition, two types of behavioral modification, the categorization of which is ambiguous (acquired olfactory preferences and shock "avoidance" learning), have been reported.

Habituation of the running response elicited by air puff stimulation of the anal cerci of the cockroach (*Periplaneta americana*) has been found by Baxter (1957), and Niklaus (1965) has shown that inhibition of firing in the giant fibers may account for a substantial portion of the behavioral habituation. Habituation and dishabituation have been demonstrated to occur in the visual interneurons in the locust tritocerebrum (Horn & Rowell, 1968; Rowell & Horn, 1968).

The few early experiments that have been adduced to demonstrate classical conditioning in insects (other than bees) may be faulted on the basis of inadequate control procedures. However, Nelson (1971) has reported a series of experiments that have been quite well controlled and that permit the exclusion of sensitization and pseudoconditioning effects. The CS was a sequential compound stimulus consisting of water followed by saline applied to the feet of the blowfly subjects. The UCS, a sucrose droplet applied to the mouth, was paired with the second component of the CS. The classical conditioning group showed significant increase in number of CR's (proboscis extension) to both CS components but especially to the component that was most immediately paired with the UCS. The CS alone and random stimulus groups showed no progressive or sustained increase in responsiveness.

Single-choice position discrimination learning based on escape from a noxious stimulus and/or avoidance of a noxious stimulies in one arm of the maze have been demonstrated in several species of cockroach: *Blattela germanica* (Hunter, 1932), *Periplaneta americana* (Minami & Dallenbach, 1946; Freckleton & Wahlsten, 1968), *Nauphoeta cincera* (Longo, 1964). Single-choice position learning has also been demonstrated in both larval and adult grain beetles (*Tenebrio molitor*) (Alloway, 1973; Borsellino, Pieratoni, & Schietti-Cavazza, 1970) and Cherkashin, Sheiman, and Stafekhina, (1968) have reported retention of such position following metamorphosis. Because it seems likely that large-scale replacement of neural tissues occurs during metamorphosis in such holometabolous insects, these results voice "serious difficulties for theories which attempt to explain insect learning in terms of simple associations between stimuli or the formation of a chain of conditioned reflexes" (Alloway, 1972). Fruitflies (Murphy, 1967, 1969) and ants (Vowles, 1964, 1965) also are capable of simple single-choice position learning. More complex sensory discriminations have also been demonstrated in single-choice tasks with a variety of insects.

Conditional attractiveness to certain odors was first reported for parasitic wasps (*Nemeritis canescens*) by Thorpe and Jones (1937). The imprinting-like phenomenon occurs when female wasps are raised on an abnormal host; when tested as adults they tend to prefer the odor of the experimental host to that of the natural host. Unlike imprinting in precocial birds, where the effect is limited to early stages of development, wasps that are first exposed as adults show a similar effect (Thorpe, 1939a, b). Furthermore, a study by Hershberger and Smith (1967) has shown that in the case of *Drosophila* such an effect is remarkably similar to the well-established secondary reinforcement phenomenon long reported for mammals. Although further research is required to more firmly establish this analogy such an effort seems to be a particularly important one within the comparative strategy.

A technique for training leg flexion in decapitated cockroaches and locusts, first developed by Horridge (1962), has been successfully exploited by several other investigators (Disterhoft, Haggerty, & Corning, 1971; Eisenstein & Cohen, 1965; Hoyle, 1965; Kerkut *et al.*, 1970). Whether true passive avoidance learning is demonstrated by these experiments or whether it is a more primitive precursor, such as aversive inhibitory conditioning, is at the moment unclear (cf. an extensive discussion of this point in Alloway, 1973). However, it is apparent that this preparation offers exciting possibilities for establishing a multilevel, polythetic definition of the behavior modification process in a simple system. It permits the coordination of neurochemical, electrophysiological, and behavioral data.

In bees, perhaps the ultimate in invertebrate learning capacities has been obtained. Associative learning can be demonstrated to be critical in daily foraging activities and orientations, and recent research has revealed even greater complexities in their behavior. For instance, the earlier findings by von Frisch indicated that the "waggle dance" represented a "language" that communicated information about the location and amount of food (von Frisch, 1950). However, the work of von Frisch has been extended by Wenner, Johnson, and Wells, who found that the bee can communicate by other means as well and that conditioning to food odors can account for food finding (reviewed in P. H. Wells, 1973).

One of the most striking demonstrations of the bee's "intellectual" capacities was provided by the research of Mazochin-Porshnyakov (1969; also reviewed in P. H. Wells, 1973). Bees were trained to prefer one of two stimuli that were presented (such as preferring a colored triangle over a square) and were then tested for their "conceptualization" by presenting the trained figure in a different color or on a different background. Bees were able to generalize and respond to "triangularity," a capacity that is difficult to achieve in many of the typical vertebrate laboratory animals.

Arthropoda: Crustacea. Krasne's (1973) careful scrutiny of learning demonstrations indicates that most categories of response modification are observed within this broad class of animals. Briefly, there is fairly solid evidence for classical conditioning, instrumental learning, instrumental discrimination learning, turn alternation, and habituation. As pointed out by Krasne, many of these demonstrations involved behavioral change that could be adaptive in the animal's normal milieu. Generally, however, these modifications are of short duration and as previously mentioned this may be a result of the lack of involvement by the cell body. Because the general learning paradigms are demonstrable in crustaceans, it is of interest to see how they perform in some of the "higher order" capacities. Krasne (1973) has considered such phenonema and reviews data that suggest crustacea may be capable of learning with secondary reinforcement and points out correctly that "secondary reinforcement is crucial to the liberation of behavior from primary reward and is therefore essential to most of the complex learned behaviors of intelligent animals" (p. 103).

"Learning to learn" is also considered to be a manifestation of higher order capacities and a simple way of assessing whether an animal has the rudiments of this ability is to subject the animal to a reversal learning task. Morrow and Smithson (1969) report evidence of this in *Porcellio* but Krasne has expressed reservations about the data because of the lack of proper controls. The Morrow and Smithson data are suggestive, however. Animals were trained to enter one arm of a dry T maze and were run to a criterion of nine out of ten correct choices. Reinforcement for a correct choice consisted of wetting the paper towel on the floor of the maze; entering an incorrect arm resulted in no reinforcement. When criterion was achieved, the task was reversed—entering the previously unrewarded arm now obtained the water reinforcement. Examination of the errors for each subject indicates an average decrease.

Crustacea may be capable of demonstrating other "mammalian" properties—certainly the "frustration" exhibited by a crab during a partial reinforcement experiment (see Krasne, 1973) and the work of Hazlett (1971), suggesting that early experience is critical for the development of certain behavioral capacities, are research findings that require considerable extension.

Mollusks. This phylum contains some behaviorally fascinating animals. The advantages of certain species in neurophysiological investigations have been known for several years. The capacities for learning that are possible have been best explored in the octopus (see Mackintosh, 1965; Sanders, 1975; M. J. Wells, 1968, for reviews). Discriminations between a wide range of visual cues are possible. Using shock as reinforcement, subjects can learn, for example, to visually distinguish between horizontal and vertical lines, triangles and circles, and triangles and squares. Texture discriminations are possible although it is the total number of sensory cells that are activated that is central in the discrimination—animals fail to discriminate between longitudinal and horizontal grooves on cylinders when the grooved proportions are comparable.

Although the octopus apparently can make adjustment according to the weight of objects that it picks up, attempts to train these animals to make weight discriminations have failed. M. J. Wells (1965) believes this inability to make use of the information concerning bodily movements results from a failure of this class of input to reach those areas of the brain critical in learning.

The octopus also demonstrates the ratlike "overtraining reversal effect." Animals learn to reverse a discrimination more readily after they have been overtrained in the original task, presumably because during the overtraining they learn to turn off other irrelevant input channels. If the animal does not learn to disregard these irrelevant inputs then distraction is more probable. The octopus has been shown to demonstrate this effect but only when the relevant cue is not an overly dominant input (M. J. Wells, 1965). In more difficult discriminations, therefore, where more inputs are being used, the animal can learn to not attend to the irrelevant. Some evidence suggesting a "learning to learn" phenomenon in octopus has also appeared.

Among the gastropod molluscs, learning demonstrations are more difficult to achieve beyond the habituation paradigms. Lee (1969) has shown that operant training is possible in *Aplysia* and Lickey (1968) has reported a conditioned discrimination between edible and inedible objects in the same animal. Stephens and McGaugh (1972) have found that the land snail (*Helix aspersa*) can be trained to inhibit its upward climbing tendency on a glass rod if a thread soaked in quinine is placed at the top. Retention of this aversive experience can persist for a week.

Long-term (3 weeks) retention of a habituated defensive reflex has been reported (Carew, Pinsker, & Kandel, 1972). Furthermore, spaced habituation trials were found to be more efficacious for producing long-term retention than massed trials.

CONCLUSIONS AND SPECULATIONS

Comparisons

A cursory review of invertebrate physiology and behavior should teach the vertebrate researcher a little humility. The emphasis of this volume has been on vertebrate behavior and neuroanatomy and the plethora of information concerned with the anatomy of various vertebrate brains and particularly, the neocortex, gives the impression that investigators are really convinced that the structures and the capacities are related. Our anthropocentrism naturally leads us to believe that somehow there is a unique meaning to our anatomy and hence there is uniqueness in our capacities and existence. However, as we have tried to indicate in this chapter, many invertebrates, in spite of the vast number of different anatomical characteristics, have achieved some of the same general functions, which again emphasizes that there have been several phylogenetic routes to similar end states. This is not to say that there is overall functional equivalence between the invertebrate and the vertebrate; the question is whether anatomy alone can tell much about the logical operations that permit "intelligence." If a few thousand cells, arranged in a completely nonvertebrate manner, can achieve functionally equivalent categories then perhaps limited and simple-minded questions are being asked of the vertebrate brain and/or the critical properties of this brain that lead to intelligent operations are not being researched. The humility lesson then, may derive from comparative considerations where similar functions are seen to be subserved by dissimilar anatomies. The vertebrate anatomy presented in these volumes has been impressive but what it has to do with the operation of vertebrate intelligent mechanisms remains obscure. Because there has been considerable divergence in substrate and considerable convergence with respect to function, what the major contribution of a comparative strategy may be should again be stressed: minimally, comparative anatomical and functional data should prevent facile conclusions concerning the unique relationships between substrate and operation. Concerns over phylogenetic origins and proximities are unnecessary when substrate–function relationships are compared in various animals.

There are "universal" mechanisms that can be uncovered in the comparative approach and that may represent some sort of phylogenetic continuity. Ionic diffusion and transport mechanisms, synaptic pharmacological operations, and many other processes can be studied in invertebrate preparations and demonstrated to be very close to the vertebrate system. However, such "universals" may be restricted to biochemical and biophysical levels. Beyond these, similarity in process should be treated more as an analog than as a phylogenetic precursor. For example, we have shown that the criteria for habituation and the experimental operations that produce habituation work for the protozoan as well as for the cat; however, one is only the model for the other because multilevel considerations indicate mechanistic dissimilarities. If interest is in the chemical manipulation of the habituation process, then different strategies probably must apply depending on multilevel characteristics. Evolution resulted in system elaboration and variation in how these biochemical and biophysical mechanisms are put to use depends on the specific system information processing characteristics. Other justifications for the comparative approach have been discussed elsewhere (Corning & Lahue, 1972).

Invertebrate–Vertebrate Intelligence Differences

Complex adaptive transactions have been achieved by the invertebrate and in many cases it can be convincingly shown that experiential events (learning) are critical in the formation of these transactions. The "routes to intelligence" that can be adopted have been discussed in the introductory section. Are there any reasonable guesses as to what distinguishes the invertebrate from the vertebrate in achieving intelligent action? We have demonstrated that examples of general functional equivalence can be found the difference between providing an example and establishing a general case should be stressed. Bees, for instance, have shown what can be interpreted as "concept" formation; the octopus is said to retain information up to a year[8]—these examples may not represent general

[8]Comment made by N. S. Sutherland during a discussion of this presentation at the conference on which this volume is based.

rules for the invertebrate but are examples of the apex in invertebrate intelligence development. There is most likely considerable overlap between various groups of invertebrates and vertebrates with respect to intelligent capacities. One question should be, how widespread or general is the capacity in one group as compared with another?

As for possible specific differences, we should like to consider two: epigenesis and suprastructural information processing. We suggest that with regard to the former, the vertebrate brain appears to be more dependent on experiential input for its structural elaboration. The adaptiveness of this epigenetic strategy is obvious: besides the facilitation of neuronal growth, glial density, circulation, etc., by experiences, it is probable that logical operations can be stipulated by these experiences. The organism is limited by its ancestry with respect to capacity but selective elaboration of cellular processes is dependent on the particular environment in which it finds itself. The degree of epigenesis varies among vertebrate groups: in cats, the deprivation of experience with "verticalness" results in a reduction of vertical detector units in the brain (Hirsch, 1972) but in rabbits, experiential deprivation does not have a similar effect (Ranney Mize, & Murphy, 1973). The influence of experience on structural elaboration may occur quite early in development and then decrease rapidly or it may be more prolonged and dependent on "critical periods" of development in various systems of the organism. In invertebrates, there is suspicion of considerable reliance on genetic preprogramming and that epigenesis is restricted to early stages of development. There is a need for "early experience" studies in invertebrates. For example, does the deprivation of rapid movement stimuli early in development lead to a reduction of units that respond to these events in adult crustaceans? Some examples of possible early experience effects have appeared (Hazlett, 1971; Thorpe, 1939a,b).

The epigenetic route to intelligence could still confine the organism to specific circuit arrangements for its adaptive transactions. Freedom from these constraints has probably been achieved in the mammalian brain by suprastructural information processing mechanisms. The macropotential activity, representing participation by ensembles of neurons in common modes of activity, could add remarkable power to the ability of an organism to encode information and integrate it with past and anticipated events. This process most likely requires certain critical numbers of neurons with a sufficient array of potential interconnections (and perhaps with particular anatomical configurations), but beyond the "critical mass," the capacities are totally dependent on the individual's experience and more independent of the structure. Circuit redundancy may permit the invertebrate to achieve maximal benefits from the "switchboard" strategy, but the critical limitation remains, i.e., the structural constraints.

This volume has emphasized structure and capacity in the vertebrate brain. These emphases may have missed the intelligence strategy that has permitted the vertebrate to escape from the anatomical confinements of evolution. Perhaps, this "token" invertebrate presentation will help to remind neuroscientists that purely connectionistic assumptions may impose serious limitations on what can be discovered about the intelligent capacities of vertebrates.

ACKNOWLEDGMENTS

Preparation of this paper and portions of the research that is reported have been supported by National Research Council Grant AO351. Portions of this paper are published elsewhere (Corning, Dyal, & Willows, 1975).

REFERENCES

Abraham, F. D., Palka, J., Peeke, H. V. S., & Willows, A. O. D. Model neural systems for the neurobiology of learning. *Behavioral Biology,* 1972, **7,** 1–24.

Adametz, J. H. Rate of recovery of functioning in cats with rostral reticular lesions. *Journal of Neurosurgery,* 1959, **16,** 85–97.

Adey, W. R. Slow electrical phenomena in the central nervous system. *Neuroscience Research Program Bulletin,* 1969, **7,** 75–180.

Agranoff, B. W. Agents that block memory. In G. C. Quarton, T. Melnechuk, & F. O. Schmitt (Eds.), *The neurosciences.* New York: Rockefeller University Press, 1967.

Alloway, T. M. Retention of learning through metamorphosis in the grain beetle (*Tenebrio molitor*) *American Zoologist,* 1972, **12,** 471–477.

Alloway, T. M. Learning in insects except Apoidea. In W. C. Corning, J. A. Dyal, & A. O. D. Willows (Eds.), *Invertebrate Learning,* Vol. II. New York: Plenum Press, 1973.

Appel, S. H., Davis, W., & Scott, S. Brain polysomes: response to environmental stimulation. *Science,* 1967, **157,** 836–838.

Applewhite, P. Temperature and habituation in a protozoan. *Nature* (London), 1968, **219,** 91–92.

Applewhite, P. The flow of ions in learning and memory. *Journal of Theoretical Biology,* 1972, **36,** 419–423.

Applewhite, P., & Gardner, F. T. Tube-escape behavior in paramecia. *Behavioral Biology,* 1973, **9,** 245–248.

Applewhite, P., & Morowitz, H. J. The micrometazoa as model systems for studying the physiology of memory. *Yale Journal of Biology & Medicine,* 1966, **39,** 90–105.

Applewhite, P., & Morowitz, H. J. Memory and the microinvertebrates. In W. C. Corning & S. C. Ratner (Eds.), *Chemistry of Learning.* New York: Plenum Press, 1967.

Applewhite, P., Gardner, F. T., & Lapan, E. Physiology of habituation learning in a protozoan. *Transactions of the New York Academy of Science,* 1969, **31,** 842–849.

Applewhite, P. B., Gardner, F., Foley, D., & Clendenin, M. Failure to condition *Tetrahymena. Scandinavian Journal of Psychology,* 1971, **12,** 65–67.

Arbit, J. A failure to confirm "latent learning in earthworms." *Worm Runner's Digest,* 1961, **3,** 129–134.

Baltzer, F. Beitrage zur Sinnesphysiologie and Psychologie der Webespinnen. *Mitteilungen der Naturforschenden Gesellschaft in Bern,* 1923, 163–187.

Bartels, M. Sinnesphysiologische und Psychologische Untersuchungen an der Trichterspinne *Agelena labyrinthica* (cl.). *Zeitschrift feur Vergluchendi Physiologie,* 1929, **10,** 527–593.

Bartels, M., & Baltzer, F. Ueber Orientierung and Gedachtnis der Netzspinne *Agelena labyrinthica* (cl.). *Revue Suisse de Zoologie,* 1928, **35,** 247–258.

Bays, S. M. Training possibilities of *Araneus diadematus. Experientia,* 1962, **18,** 423.

Baxter, C. Habituation of the roach to puffs of air. *Anatomical Record,* 1957, **128,** 521 (abstract).

Bergstrom, S. R. Acquisition of an avoidance reaction to light in the protozoa *Tetrahymena. Scandinavian Journal of Psychology,* 1968, **9,** 215–219.

Beritoff, J. S. *Vertebrate memory: Characteristics and origin.* New York: Plenum Press, 1971.

Best, J. B. Behaviour of planaria in instrumental learning paradigms. *Animal Behavior,* 1965, **13** (Suppl. 1), 69–75.

Bitterman, M. E. Phyletic differences in learning, *American Psychology,* 1965, **15,** 709–712.

Bocharova, L. S., Borovyagin, V. L., Dyakonova, T. L., Warton, S. S., & Verprintsev, B. N. Ultrastructure and RNA synthesis in a molluscan giant neuron under electrical stimulation. *Brain Research,* 1972, **36,** 371–384.

Borsellino, A., Pieratoni, R., & Schietti-Cavazza, B. Survival in adult mealworm beetles (*Tenebrio molitor*) of learning acquired at the larval stage. *Nature* (London), 1970, **225,** 963–964.

Bramstedt, F. Dressurversuche mit *Paramecium caudatum* und *Stylonychia mytilus. Zeitschrift für Vergleichende Physiologie,* 1935, **22,** 490–516.

Bruner, J., & Kennedy, D. Habituation: occurrence at a neuromuscular junction. *Science,* 1970, **169,** 92–94.

Bullock, T. H., & Horridge, G. A. *Structure and Function in the Nervous Systems of Invertebrates,* San Francisco: Freeman, 1965.

Carew, T. J., Pinsker, H. M., & Kandel, E. R. Long-term habituation of a defensive withdrawal reflex in *Aplysia. Science,* 1972, **175,** 451–454.

Castellucci, V., Pinsker, H., Kupfermann, I., & Kandel, E. Neuronal mechanisms of habituation and dishabituation of the gill-withdrawal reflex in *Aplysia. Science,* 1969, **167,** 1745–1748.

Cherkashin, A. N., Sheiman, E. M., & Stafekhina, V. S. O sokhanenii vremennikh svyazei oo nasekhomikh v protsesse metamorfoza. (Conservation of conditioned reflexes during metamorphosis in insects.) *Zhurnal Evolutsionnoi Biochimii i Fiziologii,* 1968, 117–120.

Clark, R. B. Habituation of the polychaete *Nereis* to sudden stimuli. I. General properties of the habituation process. *Animal Behavior,* 1960, **8,** 82–91. (a)

Clark, R. B. Habituation of the polychaete *Nereis* to sudden stimuli. II. The biological significance of habituation. *Animal Behavior,* 1960, **8,** 92–103. (b)

Clark, R. B. The integrative action of a worm's brain. *Symposia of the Society of Experimental Biology,* 1965, **20,** 345–379.

Cohen, M. J. Some cellular correlates of behavior controlled by an insect ganglion. In W. C. Corning, & S. C. Ratner (Eds.), *Chemistry of learning,* New York: Plenum Press, 1967.

Cohen, M. J. A comparison of invertebrate and vertebrate central neurons. In F. O. Schmitt (Ed.), *The neurosciences, second study program,* New York: Rockefeller University Press, 1972.

Cook, A. Habituation in a freshwater snail (*Limnaea stagnalis*). *Animal Behavior,* 1971, **19,** 463–574.

Corning, W. C. Retention of a position discrimination after regeneration in planarians. *Psychonomic Science,* 1966, **5,** 17–18.

Corning, W. C. Regeneration and retention of acquired information. In W. C. Corning & S. C. Ratner (Eds.), *Chemistry of learning.* New York: Plenum Press, 1967.

Corning, W. C. Recent learning demonstrations and some biochemical correlates in planarians and protozoans. In E. Adam (Ed.), *Biology of memory.* New York: Plenum Press, 1971. (a)

Corning, W. C. Conditioning and "transfer" of training in a colonial ciliate: A summary of the work of N. N. Plavilstchikov. *Journal of Biological Psychology,* 1971, **13,** 39–41. (b)

Corning, W. C., Dyal, J. A., & Willows, A. O. D. *Invertebrate Learning,* Vol. III. in preparation, 1975.

Corning, W. C., & John, E. R. Effect of ribonuclease on retention of conditioned response in regenerated planarians. *Science,* 1961, **134,** 1363–1364.

Corning, W. C., & Kelly, S. Platyhelminthes: The Turbellarians. In W. C. Corning, J. A. Dyal, & A. O. D. Willows (Eds.), *Invertebrate learning,* Vol. I. New York: Plenum Press, 1973.

Corning, W. C., & Lahue, R. Reflex "training" in frogs. *Psychonomic Science,* 1971, **23,** 119–120.

Corning, W. C., & Lahue, R. Invertebrate strategies in comparative learning studies. *American Zoologist,* 1972, **12,** 455–469.

Corning, W. C., & Ratner, S. C. *Chemistry of learning: Invertebrate research,* New York: Plenum Press, 1967.

Corning, W. C., & Riccio, D. The planarian controversy. In W. Byrne (Ed.), *Molecular approaches to learning and memory.* New York: Academic Press, 1970.

Corning, W. C., & Von Burg, R. Behavioral and neurophysiological investigation of *Limulus polyphemus.* In J. Salanki (Ed.), *Neurobiology of invertebrates.* New York: Plenum Press, 1968.

Corning, W. C., & Von Burg, R. Protozoa. In W. C. Corning, J. A. Dyal, & A. O. D. Willows (Eds.), *Invertebrate learning,* Vol. I. New York: Plenum Press, 1973.

Crawford, F. T., & Skeen, L. C. Operant responding in the planarian: a replication study. *Psychological Reports,* 1967, **20,** 1023–1027.

Datta, L. E. Some experiments on learning in the earthworm, *Lumbricus terrestris. American Journal of Psychology,* 1962, **75,** 531–553.

Disterhoft, J. F., Haggerty, R., & Corning, W. C. An analysis of leg position learning in the cockroach yoked control. *Physiology of Behavior,* 1971, **7,** 1026–1029.

Dyal, J. A. Transfer of behavioral bias: reality and specificity. In E. J. Fjerdingstad (Ed.), *Chemical transfer of learned information.* New York: American Elsevier, 1971.

Dyal, J. A. Behavior modification in annelids. In W. C. Corning, J. A. Dyal, & A. O. D. Willows (Eds.), *Invertebrate learning,* Vol. I. New York: Plenum Press, 1973.

Dyal, J. A., & Corning, W. C. Invertebrate learning and behavior taxonomies. In W. C. Corning, J. A. Dyal, & A. O. D. Willows (Eds.), *Invertebrate learning,* Vol. I. New York: Plenum Press, 1973.

Dyal, J. A., & Hetherington, K. Habituation in the polychaete, *Hesperonoe adventor. Psychonomic Science,* 1968, **13,** 263–264.

Eisenstein, E. M., & Cohen, M. J. Learning in an isolated insect ganglion. *Animal Behavior,* 1965, **13,** 104–108.

Eisenstein, E. M., & Peretz, B. Comparative aspects of habituation. In M. J. Herz & H. V. S. Peeke (Eds.), *Habituation: Behavioral studies and physiological substrates.* New York: Academic Press, 1973.

Evans, S. M. Behaviour of the polychaete *Nereis* in T-mazes. *Animal Behavior,* 1963, **11,** 379–392.

Evans, S. M. Habituation of the withdrawal response in nereid polychaetes I. The habituation process in *Nereis diversicolor. Biological Bulletin,* 1969, **137,** 105–117.

Fisher, A. E. Maternal and sexual behavior induced by intracranial chemical stimulation. *Science,* 1956, **124,** 228–229.

Fjerdingstad, E. J. *Chemical transfer of learned information.* New York: American Elsevier, 1971.

Flavell, J. H. *The developmental psychology of Jean Piaget.* New York: Van Nostrand, 1963.

Flint, P. The effect of sensory deprivation on the behavior of the polychaete *Nereis* in T-mazes. *Animal Behavior,* 1965, **13,** 187–193.

Franzisket, L. Characteristics of instinctive behavior and learning in reflex activity of the frog. *Animal Behavior,* 1963, **11,** 318–334.

Freckleton, W. C., & Wahlsten, D. Carbon dioxide induced amnesia in the cockroach. *Psychonomic Science,* 1968, **12,** 179–180.

French, J. W. Trial-and-error learning in *Paramecium. Journal of Experimental Psychology,* 1940, **26,** 609–613.

Gaito, J. *Macromolecules and behavior* (2nd ed.) New York: Appleton-Century-Crofts, 1972.

Gardner, L. E. Retention and overhabituation of a dual-component response in *Lumbricus terrestris. Journal of Comparative and Physiological Psychology,* 1968, **66,** 315–318.

Gee, W. The behavior of leeches with especial reference to its modifiability. *University of California Publications in Zoology,* 1913, **11,** 197–305.

Gelber, B. Investigations of the behavior of *Paramecium aurelia:* I. Modification of behavior after training with reinforcement. *Journal of Comparative and Physiological Psychology,* 1952, **45,** 58–65.

Gelber, B. Studies of the behavior of *Paramecium aurelia. Animal Behavior,* 1965, **13** (Suppl. 1), 21–29.

Grabowski, U. Experimentelle Untersuchungen ueber das angebliche Lernervermoegen von *Paramecium. Zeitschrift feur Tierpsychologie,* 1939, **2,** 265–282.

Groves, P. M., & Thompson, R. F. Habituation: a dual-process theory. *Psychological Reviews,* 1970, **77,** 419–450.

Haigler, H. J., & Von Baumgarten, R. J. Facilitation of excitatory post-synaptic potentials in the giant cell in the left pleural ganglion of *Aplysia californica. Comparative Biochemistry of Physiology,* 1972, **41A,** 7–16.

Hanzel, T. E., & Rucker, W. Escape training in paramecia. *Journal of Biological Psychology,* 1971, **13,** 24–28.

Harden, C. M. Behavioral modification of *Stentor coeruleus.* Reprinted in W. C. Corning, J. A. Dyal, & A. O. D. Willows (Eds.), *Invertebrate learning,* Vol. I. New York: Plenum Press, 1973.

Hazlett, E. A. Influence of rearing conditions on initial shell entering behavior of a hermit crab (Decapoda Paguridea). *Crustaceana,* 1971, **20,** 167–170.

Held, R. M. Plasticity in sensory-motor systems. *Scientific American,* 1965, **220,** 84–94.

Hershberger, W. A., & Smith, M. P. Conditioning in *Drosophila melanogaster. Animal Behavior,* 1967, **15,** 259–262.

Hess, W. R. *The functional organization of the diencephalon.* New York: Grune and Stratton, 1957.

Hirsch, H. V. B. Role of function in the development and maintenance of the cat visual system. *Neurosciences Research Program Bulletin,* 1972, **10,** 291–293.

Holmgren, B., & Frenk, S. Inhibitory phenomena and habituation at the neuronal level. *Nature* (London), 1961, **192,** 1294–1295.

Horn, G., & Rowell, C. H. F. Medium and long-term changes in the behavior of visual neurones in the tritocerebrum of locusts. *Journal of Experimental Biology,* 1968, **49,** 143–169.

Horridge, G. A. Analysis of the rapid responses of *Nereis* and *Harmothoe* (Annelida). *Proceedings of the Royal Society (London) Series E,* 1959, **150,** 245–262.

Horridge, G. A. Learning leg position by the ventral nerve cord in headless insects. *Proceedings of the Royal Society, Series B.,* 1962, **157,** 33–52.

Horridge, G. A. *Interneurons.* San Francisco: Freeman, 1968.

Hoyle, G. Neurophysiological studies on "Learning" in headless insects. In J. E. Treherne & J. W. L. Beament, (Eds.), *The physiology of the insect central nervous system.* New York: Academic Press, 1965.

Hoyle, G. Cellular mechanisms underlying behavior–neuroethology. *Advances in Insect Physiology,* 1970, **12,** 349–444.

Hubel, D. H., & Wiesel, T. N. Receptive fields of cells in striate cortex of very young, visually inexperienced kittens. *Journal of Neurophysiology,* 1963, **26,** 994–1002.

Hunter, W. S. The effect of inactivity produced by cold upon learning and retention in the cockroach, *Blattella germanica. Journal of Genetic Psychology,* 1932, **41,** 253–266.

Hyden, H. Biochemical changes accompanying learning. In G. C. Quarton, T. Melnechuk, & F. O. Schmitt (Eds.), *The neurosciences.* New York: Rockefeller University Press, 1967.

Jacobson, A. L. Classical conditioning and the planarian. In W. C. Corning & S. C. Ratner, (Eds.), *Chemistry of learning.* New York: Phenum Press, 1967.

Jacobson, A. L. Progress in the study of learning and chemical transfer in planarians. In E. J. Ejerdingstad (Ed.), *Chemical transfer of learned information.* New York: American Elsevier, 1971.

Jensen, D. D. Experiments on "learning" in paramecia. *Science,* 1957a, **125,** 191–192.

Jensen, D. D. More on "learning" in paramecia. *Science,* 1957b, **126,** 1341–1342.

Jensen, D. D. Polythetic operationism and the phylogeny of learning. In W. C. Corning & S. C. Ratner (Eds.), *Chemistry of learning.* New York: Plenum Press, 1967.

John, E. R. *Mechanisms of memory.* New York: Academic Press, 1967.

John, E. R. Switchboard versus statistical theories of learning and memory. *Science,* 1972, **177,** 850–864.

John, E. R., & Killam, K. F. Electrophysiological correlates of differential approach-avoidance conditioning in the cat. *Journal of Nervous & Mental Disease,* 1960, **131,** 183.

Kandel, E. R., & Tauc, L. Mechanisms of heterosynaptic facilitation in the giant cell of the abdominal ganglion of *Aplysia depilans. Journal of Physiology,* 1965, **181,** 28–47.

Kandel, E. R., & Spencer, W. A. Cellular neurophysiological approaches in the study of learning. *Physiological Review,* 1968, **48,** 65–134.

Katz, M., & Deterline. W. A. Apparent learning in the paramecium. *Journal of Comparative and Physiological Psychology,* 1958, **51,** 243–247.

Kennedy, D., Selverston, A. I., & Remler, M. P. Analysis of restricted neural networks. *Science,* 1969, **164,** 1488–1495.

Kerkut, G. A., Oliver, G. W. C., Rick, J. T., & Walker, R. J. The effects of drugs on learning in a simple preparation. *Comparative & General Pharmacology,* 1970, **1,** 437–483.

Kinastowski, W. Der Einfluss der mechanischen Reize auf die Kontraktilitat von *Spirostomum ambiguum. Acta Protozoologica,* 1963, **1,** 201–222. (a)

Kinastowski, W. Das Problem "des lernes" bei *Spriostomum ambiguum. Acta Protozoologica,* 1963, **1,** 223–236. (b)

Krasne, F. B. Escape from recurring tactile stimulation in *Branchiomma vesiculosum. Journal of Experimental Biology,* 1965, **42,** 307–322.

Krasne, F. B. Excitation and habituation of the crayfish escape reflex: The depolarizing response in lateral giant fibres of the isolated abdomen. *Journal of Experimental Biology,* 1969, **50,** 29–46.

Krasne, F. B. Learning in Crustacea. In W. C. Corning, J. A. Dyal, & A. O. D. Willows (Eds.), *Invertebrate learning,* Vol. II. New York: Plenum Press, 1973.

Krasne, F. B., & Roberts, A. Habituation of crayfish escape response during release from inhibition induced by picrotoxin. *Nature (London),* 1967, **215,** 769–770.

Krasne, F. B., & Woodsmall, K. S. Waning of the crayfish escape response as a result of repeated stimulation. *Animal Behavior,* 1969, **17,** 416–424.

Kristan, W. B. Jr. Plasticity of firing patterns in neurons of *Aplysia* pleural ganglion. *Journal of Neurophysiology,* 1971, **34,** 321–336.

Kuenzer, P. P. Verhaltenphysiologische Untersuchungen uber das Zucken des Regenwurms. *Zeitschrift fuer Tierpsychologie,* 1958, **15,** 31–49.

Lahue, R. The Chelicorates. In W. C. Corning, J. A. Dyal, & A. Q. D. Willows (Eds.), *Invertebrate learning,* Vol. II. New York: Plenum Press, 1973. (a)

Lahue, R. "Mechanisms of habituation in the abdominal ganglia of *Limulus polyphemus.*" Unpublished doctoral dissertation, University of Waterloo, Waterloo, Ontario, Canada, 1973. (b)

Lahue, R., & Corning, W. C. Habituation in *Limulus* abdominal ganglia. *Biological Bulletin,* 1971, **140,** 427–439. (a)

Lahue, R., & Corning, W. C. Habituation in *Limulus* abdonimal ganglia: an exercise in paleopsychology. *Canadian Psychologist,* 1971, **12** (Suppl. 2), 193–194. (b)

Lahue, R., & Corning. W. C. Incremental and decremental processes in *Limulus* ganglia: Stimulus frequency and ganglion organization influences. *Behavioral Biology,* 1973, 8, 637–653.

Lashley, K. S. In search of the engram. *Symposia of the Society of Experimental Biology,* 1950, **4,** 454–482.

Lee, R. M. Conditioning of a free operant response in planaria. *Science,* 1963, **139,** 1048–1049.

Lee, R. M. Effects of contingent water level variation. *Communications in Behavioral Biology,* 1969, **3,** 157–164.

Lettvin, J. Y., Maturana, H. R., McCulloch, W. S., & Pitts, W. H. What the frog's eye tells the frog's brain. *Proceedings of the Institute of Radio Engineers,* 1959, **47,** 1940–1951.

Levi-Montalcini, R., & Chen, J. S. *In vitro* studies of the insect embryonic nervous system. In S. H. Barondes (Ed.), *Cellular dynamics of the neuron.* New York: Academic Press, 1969.

Lickey, M. E. Learned behavior in *Aplysia vaccaria. Journal of Comparative and Physiological Psychology,* 1968, **66,** 712–718.

Longo, N. Probability learning and habit reversal in the cockroach. *American Journal of Psychology,* 1964, **77,** 29–41.

Luco, J. V. A study of memory in insects. In F. F. Kao, K. Koizumi, & M. Vasalle (Eds.), *Research in physiology.* Bologna: Aulo Gaggi, 1971.

Luria, A. R. *Higher cortical functions in man.* New York: Basic Books, 1966.

Mackintosh, N. J. Discrimination learning in the octopus. *Animal Behavior,* 1965, **13** (Suppl. 1), 129–134.

Makous, W. L. Conditioning in the horseshoe crab. *Psychonomic Science,* 1969, **14,** 4–6.

Mazochin-Porshnyakov, G. A. Die Fahigkeit der Bienen, visuelle Reize zu generalisieren. *Zeitschrift fuer Vergleichende Physiologie,* 1969, **65,** 15–28.

McConnell, J. V. Memory transfer through cannibalism in planarians. *Journal of Neuropsychiatry,* 1962, **3** (Suppl. 1), 542.

McConnell, J. V. Comparative physiology: learning in invertebrates. *Annual Review of Physiology,* 1966, **28,** 107–136.

McConnell, J. V., Jacobson, A. L., & Kimble, D. P. The effects of regeneration upon retention of a conditioned response in the planarian. *Journal of Comparative Physiology & Psychology,* 1959, **52,** 1–5.

Mekler, L. B. Mechanism of biological memory. *Nature* (London), 1967, **215,** 481–484.

Minami, H., & Dallenbach, K. M. The effect of activity upon learning and retention in the cockroach, *Periplaneta americana. American Journal of Psychology,* 1946, **59,** 1–58.

Morrell, R. Effect of anodal polarization on the firing pattern of single cortical cells. *Annals of the New York Academy of Science,* 1961, **92,** 860–876.

Morrow, J. E., & Smithson, E. L. Learning sets in an invertebrate. *Science,* 1969, **164,** 850–851.

Moruzzi, G., & Magoun, H. W. Brain stem reticular formation and activation of the EEG. *Electroencephalography & Clinical Neurophysiology,* 1949, **1,** 455–473.

Murphy, R. M. Instrumental conditioning of the fruit fly, *Drosophila melanogaster. Animal Behavior,* 1967, **15,** 153–161.

Murphy, R. M. Spatial discrimination performance of *Drosophila melanogaster:* Some controlled and uncontrolled correlates. *Animal Behavior,* 1969, **17,** 43–46.

Nelson, M. C. Classical conditioning in the blowfly (*Phormia regina*): Associative and excitatory factors. *Journal of Comparative Physiological Psychology,* 1971, **77,** 353–368.

Niklaus, R. Die erregung einzelner fadenhaare von *Periplaneta americana* in abhangikeit von der grosse und richtung der anstenkung. *Zeitschrift fuer Vergleichende Physiologie,* 1965, **50,** 331–362.

Olds, J. Mechanisms of instrumental conditioning. In R. Hernandez-Peon (Ed.), *The physiological Basis of mental activity.* New York: Elsevier, 1963.

Palka, J. Discrimination between movements of eye and object by visual interneurones of crickets. *Journal of Experimental Biology,* 1969, **50,** 723–732.

Peeke, H. V. S., Herz, M. J., & Wyers, E. J. Forward conditioning, backward conditioning and pseudoconditioning sensitization in the earthworm (*Lumbricus terrestris*). *Journal of Comparative and Physiological Psychology,* 1967, **53,** 534–536.

Peters, H. Experimente uber die Orientierung der Kreuzspinne *Epeira diademata* Cl. im Netz. *Zoologisches Jahrbuch Abteilungen,* 1932, **51,** 239–288.

Pinsker, H., Kupferman, I., Castellucci, V., & Kandel, E. Habituation and dishabituation of the gill-withdrawal reflex in *Aplysia. Science,* 1969, **167,** 1740–1742.

Plavilstchikov, N. N. Observations sur L'excitabilite des infusories. *Russki Arkiv Prostita,* 1928, 7, 1–24.

Pumphrey, R. J., & Rawdon-Smith, A. F. Synaptic transmission of nerve impulses through the last abdominal ganglion of the cockroach. *Proceedings of the Royal Society (London), Series E,* 1939, **122,** 106–118.

Ranney Mize, R., & Murphy, E. H. Selective visual experience fails to modify receptive field properties of rabbit striate-cortex neurons. *Science,* 1973, **180,** 320–323.

Rasch, E., Swift, H., Riesen, A. H., & Chow, K. L. Altered structure and composition of retinal cells in dark reared mammals. *Experimental Cell Research,* 1961, **25,** 348–363.

Ratner, S. C. Habituation and retention of habituation in the leech (*Macrobdella decora*). *Journal of Comparative and Physiological Psychology,* 1972, **81,** 115–121.

Ratner, S. C., & Miller, K. R. Effects of spacing of training and ganglia removal on conditioning in earthworms. *Journal of Comparative Physiology & Psychology,* 1959, **52,** 102–105.

Ray, A. J. Instrumental light avoidance by the earthworm. *Communications in Behavioral Biology,* 1968, **1,** 205–208.

Razran, E. *Mind in evolution–An east–west synthesis of learned behavior and cognition.* Boston: Houghton-Mifflin, 1971.

Roberts, M. B. V. The giant fibre reflex of the earthworm, *Lumbricus terrestris* L. I. The rapid response. *Journal of Experimental Biology,* 1962, **39,** 219–227.

Rosenzweig, M. R., Bennett, E. L., & Diamond, M. C. In J. Gaito (Eds.), *Macromolecules and behavior.* (2nd ed.) New York: Appleton-Century-Crofts, 1972.

Ross, D. M. The behavior of sessile coelenterates in relation to some conditioning experiments. *Animal Behavior,* 1965, **13,** (Suppl. 1), 43–55.

Rowell, C. H. F. Variable responsiveness of a visual interneurone in the free-moving locust, and its relation to behavior and arousal. *Journal of Experimental Biology,* 1971, **55,** 727–747.

Rowell, C. H. F., & Horn, G. Dishabituation and arousal in the response of single nerve cells in an insect brain. *Journal of Experimental Biology,* 1968, **49,** 171–183.

Rushforth, N. B. Behavioral studies of the coelenterate *Hydra pirardi* Brien. *Animal Behavior,* 1965, **13** (Suppl. 1), 30–42.

Rushforth, N. B. Chemical and physical factors affecting behavior in *Hydra.* In W. C. Corning & S. C. Ratner (Eds.), *Chemistry of learning.* New York: Plenum Press, 1967.

Rushforth, N. B. Behavioral modification in coelenterate learning. In W. C. Corning, J. A. Dyal, & A. O. D. Willows (Eds.), *Invertebrate learning,* Vol. I. New York: Plenum Press, 1973.

Rushforth, N. B., Burnett, A. L., & Maynard, R. Behavior in *Hydra.* Contraction responses of *Hydra pirardi* to mechanical and light stimuli. *Science,* 1963, **139,** 760–761.

Rusinov, V. S. An electrophysiological analysis of the connection function in the cerebral cortex in the presence of a dominant focus. *Abstracts of the Proceedings of the 19th International Congress of Physiology and Science,* 1953, Montreal, 147–151.

Sanders, G. D. Octopus learning. In W. C. Corning, J. A. Dyal, & A. O. D. Willows (Eds.), *Invertebrate learning,* Vol. III. New York: Plenum Press, 1975.

Schapiro, S., & Vukovich, K. R. Early experience effect upon cortical dendrites: A proposed model for development. *Science,* 1970, **167,** 292–294.

Sluckin, W. *Imprinting and early learning.* Chicago: Aldine, 1965.

Smith, J. C., & Baker, H. D. Conditioning in the horseshoe crab. *Journal of Comparative and Physiological Psychology,* 1960, **52,** 279–281.

Smith, G. E., & Dinkes, I. "Passive avoidance learning by the earthworm." Paper delivered at the Eastern Psychological Association Meeting, New York City, April 15–17, 1971.

Sokal, R. R. Numerical taxonomy. *Scientific American,* 1966, **215,** 106.

Stein, D. G., Rosen, J. J., Graziadei, J., Mishkin, D., & Brink, J. J. Central nervous system: recovery of function. *Science,* 1969, **166,** 528–530.

Stephens, G. J., & McGaugh, J. L. Biological factors related to learning in the land snail (*Helix aspersa* Muller). *Animal Behavior,* 1972, **20,** 309–315.

Szlep, R. On the plasticity of instinct of a garden spider (*Aranea diadema* L.) construction of a cobweb. *Acta Biologique Experimentalis, Polish Academy of Science,* 1952, **16,** 5–24.

Thompson, R. F., & Spencer, W. A. Habituation: a model phenomenon for the study of neuronal substrates of behavior. *Psychological Reviews,* 1966, **73,** 16–43.

Thorpe, W. H. Further studies on olfactory conditioning in a parasitic insect. The nature of the conditioning process. *Proceedings of the Royal Society of London,* 1939a, *B126,* 379–397.

Thorpe, W. H. Further studies of pre-imaginal olfactory conditioning in insects. *Proceedings of the Royal Society of London,* 1939b, **B127,** 424–433.

Thorpe, W. H., & Jones, F. G. W. Olfactory conditioning and its relation to the problem of host selection. *Proceedings of the Royal Society of London,* 1937, **B124,** 56–81.

Van Oye, P. Over het geheugen hijfr flatwormen en andere biologische wearnemingen bji deze dieran. *Natuurwetenschappelyk* Tijdschrift (Ghent), 1920, **2,** 1.

Von Burg, R., & Corning, W. C. Cardioregulatory properties of the abdominal ganglion in *Limulus. Canadian Journal of Physiology & Pharmacology,* 1970, **45,** 333–346.

von Frisch, K. *Bees, their vision, chemical senses and language.* Ithaca: Cornell University Press, 1950.

Vowles, D. M. Neural mechanisms in insect behavior. In W. H. Thorpe & O. L. Zangwill (Eds.), *Current problems in animal behavior.* Cambridge, Mass: Harvard University Press, 1961.

Vowles, D. M. Olfactory learning and brain lesions in the wood ant (*Formica rufa*). *Journal of Comparative and Physiological Psychology,* 1964, **58,** 105–111.

Vowles, D. M. Maze learning and visual discrimination in the wood ant (*Formica rufa*). *British Journal of Psychology,* 1965, **56,** 15–31.

Wasserman, G. S., & Patten, D. G. Avoidance conditioning in *Limulus. Psychonomic Science,* 1969, **15,** 143.

Wawrzynczyk, S. Badania nad pamiecia *Spirostomum ambiguum major. Acta Bioloque Experimentalis, Polish Academy of Sciences,* 1937, **11,** 57–77.

Wells, M. J. Learning and movement in octopuses. *Animal Behavior,* 1965, **13** (Suppl. 1), 115–128.

Wells, M. J. Sensitization and the evolution of associative learning. In J. Salanki (Ed.), *Neurobiology of invertebrates.* New York: Plenum Press, 1968.

Wells, P. H. Training flatworms in a Van Oye maze. In W. C. Corning & S. C. Ratner (eds.), *Chemistry of learning.* New York: Plenum Press, 1967.

Wells, P. H. Honey bees. In W. C. Corning, J. A. Dyal, & A. O. D. Willows (Eds.), *Invertebrate learning,* Vol. II. New York: Plenum Press, 1973.

Wickelgren, B. C. Habituation of spinal motoneurons. *Journal of Neurophysiology,* 1967a, **30,** 1404–1423.

Wickelgren, B. C. Habituation of spinal interneurons. *Journal of Neurophysiology,* 1967b, **30,** 1424–1438.

Wiersma, C. A. G., & Yanagisawa, K. On types of interneurons responding to visual stimulation present in the optic nerve of the rock lobster, *Panulirus interruptus. Journal of Neurobiology,* 1971, **2,** 291–309.

Willows, A. O. D. Neuronal network triggering a fixed action pattern. *Science,* 1969, **166,** 1549–1551.

Willows, A. O. D. Learning in gastroped molluscs. In W. C. Corning, J. A. Dyal, & A. O. D. Willows (Eds.), *Invertebrate learning,* Vol. II. New York: Plenum Press, 1973.

Wilson, D. L., & Berry, R. W. The effect of synaptic stimulation on RNA and protein metabolism in the R2 soma of *Aplysia. Journal of Neurobiology,* 1972, **3,** 369–379.

Wine, J. J. Invertebrate central neurons: Orthograde degeneration and retrograde changes after axotomy. *Experimental Neurology,* 1973, **38,** 157–169.

Wood, D. C. Parametric studies of the response decrement produced by mechanical stimuli in the protozoan, *Stentor coeruleus. Journal of Neurobiology,* 1970a, **1,** 345–360.

Wood, D. C. Electrophysiological studies of the protozoan, *Stentor coeruleus. Journal of Neurobiology,* 1970b, **1,** 363–377.

Wood, D. C. Electrophysiological correlates of the response decrement produced by mechanical stimuli in protozoan, *Stentor coeruleus. Journal of Neurobiology,* 1971, **2,** 1–11.

Wood, D. C. Generalization of habituation between different receptor surfaces of *Stentor. Physiology and Behavior,* 1973, **9,** 161–165.

Woodson, F. B. J., Schlapfer, W. T., & Barondes, S. H. Postural avoidance learning in the headless cockroach without detectable changes in ganglionic cholinesterase. *Brain Research,* 1972, **37,** 348–352.

Wyers, E. J., Peeke, H. V. S., & Herz, M. J. Partial reinforcement and resistance to extinction in the earthworm. *Journal of Comparative and Physiological Psychology,* 1964, **57,** 113–116.

Zelman, A., Kabat, L., Jacobson, R., & McConnell, J. V. Transfer of training through injection of "conditioned" RNA into untrained planarians. *Worm Runner's Digest,* 1963, **5,** 14–17.

Zubkov, A., & Polikarpov, G. G. Conditioned reflex in coelenterates. *Advances in Modern Biology, Moscow,* 1951, **32,** 301–302.

Zucker, R. S., Kennedy, D., & Selverston, A. I. Neuronal circuit mediating escape responses in crayfish. *Science,* 1971, **173,** 645–650.

Author Index

A

Abraham, F. D., 221, 222, *255*
Adametz, J. H., 234, *255*
Ades, H. W., 26, *43*
Adey, W. R., 35, *43*, 235, *255*
Agranoff, B. W., 240, *255*
Alexander, R. M., 119, *121*
Alloway, T. M., 248, 249, 250, *255*
Alpern, M., 100, *106*
Andrew, R. J., 195, *197*
Andrews, A. M., 172, *173*
Andy, O. J., 23, 30, *45*
Appel, S. H., 101, *106*, 217, *255*
Applewhite, P., 224, 232, 234, 242, 243, *255*
Arbit, J., 247, *255*
Ariëns-Kappers, C. U., 3, *10*, 156, *166*
Ashmole, N. P., 19, *21*
Atz, J. W., 137, *140*, 154, 155, 158, 160, 161, 162, 164, *166*, 193, 194, 195, *197*, 199, *212*

B

Bajema, C., 97, *103*
Baker, H. D., 248, *262*
Baltzer, F., 216, 239, *255*
Barghusen, H., 61, *75*
Barondes, S. H., 240, *263*
Bartels, M., 216, 239, *255*
Bauchot, R., 23, 28, 30, *43, 45*, 176, *178*
Baxter, C., 249, *256*
Bays, S. M., 248, *256*
Benjamin, B., 96, *103*
Bennett, E. L., 217, *261*
Berg, I. A., 14, *21*
Bergquist, H., 147, *150*
Bergstrom, S. R., 242, *256*
Beritoff, J. S., 216, 239, *256*
Berry, R. W., 240, *263*
Best, J. B., 245, *256*
Bitterman, M. E., *21*, 216, 223, *256*
Bloom, L. M., 211, *212*
Bocharova, L. S., 240, *256*
Bock, W. J., 132, 136, *140*, 144, *150*, 169, *173*, 189, *197*
Bodenheimer, T. S., 143, 147, 148, 149, *151*
Bodmer, W. F., 95, *104*
Böök, J. A., 96, *103*
Bok, S. T., 23, *43*
Borovyagin, V. L., 240, *256*
Borsellino, A., 249, *256*
Bowerman, M., 211, *212*
Boyden, A., 131, 136, *140*, 185, 188, *197*
Bramstedt, F., 242, *256*
Brazier, M. A. B., 23, *43*
Brink, A. S., 58, 63, *75*
Brink, J. J., 219, *262*
Broekhuysen, G. J., 15, *22*, 184, *198*
Brown, F. A., 156, 159, *167*
Brown, R., 211, *212*
Bruner, J., 227, 235, *256*
Bullock, T. H., 26, *43*, 231, 233, *256*

Burnett, A. L., 225, *262*
Bush, G. L., 103, *103*
Butler, A. B., 109, *112*

C

Cahn, P. H., 156, *166*
Campbell, C. B. G., 108, 109, 111, *112*, 134, 137, *140, 141*, 143, 144, 145, 147, *150*, 153, 154, 165, *166*, 181, 185, 186, 187, 189, 191, *197*
Campos, G. B., 41, *45*
Carew, J. J., 229, 252, *256*
Carter, C. O., 96, *104*
Caspary, D., 5, *10*
Castellucci, V., 229, *256, 261*
Cavalli-Sforza, L. L., 95, *104*
Chen, J. S., 232, *260*
Cherkashin, A. N., 249, *256*
Chomsky, C., 211, *212*
Chomsky, N., 202, *212*
Chow, K. L., 217, *261*
Clark, R. B., 226, 246, 247, *256*
Clendenin, M., 242, *255*
Coble, J. R., 100, *104, 106*
Cohen, M. J., 231, 250, *256, 257*
Coleman, W., 1, *10*
Collins, R. A., 101, *104*
Cook, A., 228, 229, *256*
Corning, W. C., 218, 221, 222, 223, 224, 227, 234, 235, 237, 238, 241, 243, 244, 245, 248, 250, 253, 255, *256, 257, 260, 262*
Cox, V. C., 39, *45*
Cracraft, J., 189, *198*
Crawford, F. T., 245, *257*
Creel, D. J., 101, *104*
Crile, G., 28, 29, *43*
Crosby, E. C., 3, *10*, 156, *166*
Crowson, R. A., 164, *166*
Cullen, E., 19, 20, *21*
Cullen, J. M., 19, *21*

D

Daanje, A., 158, *166*
Dallenbach, K. M., 249, *260*
Darwin, C., 116, *121*, 154, *166*
Datta, L. F., 247, *257*
Davis, W., 217, *255*
de Beer, G., 136, *140*, 171, *173*
DeFries, J. C., 83, 85, 90, 93, 94, 96, 101, *104, 105, 106*
Delacour, J., 19, *22*
deLuca, C., 126, *126*
Denison, R., 50, 53, *75*
Deterline, W. A., 241, *259*
Diamond, I. T., 109, 110, *112*
Diamond, M. C., 217, *261*
Dimond, S. J., 161, *166*
Dinkes, I., 246, *262*
Disterhoft, J. F., 250, *257*
Dizinno, G., 100, *106*
Dobzhansky, T., 79, 80, 82, 90, *104*
Dyakonova, T. L., 240, *256*
Dyal, J. A., 217, 218, 222, 238, 246, 255, *257*

E

Eakin, R. M., 53, *77*
Ebbesson, S. O. E., 4, *10*
Ebner, F. F., 4, *10*, 109, *112*, 150, *150*
Eckland, B. K., 97, *104*
Edinger, T., 51, 52, 53, 62, 63, 64, 67, 68, 69, *75, 76*
Ehrman, L., 83, *104*
Eisenstein, E. M., 243, 250, *257*
Eldredge, N., 121, *122*, 165, *167*
Erlenmeyer-Kimling, L., 93, *104*
Evans, S. M., 226, 246, 247, *257*

F

Falconer, D. S., 85, 94, 97, *104*
Feekes, F., 15, *22*
Feekes, V., 184, *198*
Fisher, A. E., 219, 240, *257*
Fisher, R. A., 89, *104*
Fjerdingstad, E. J., 238, *257*
Flavell, J. H., 223, *257*
Flint, P., 247, *258*
Foley, D., 242, *255*
Ford, F. B., 81, 92, 94, 103, *104*
Fraga, M. G., 126, *126*
Franzisket, L., 235, *258*
Freckleton, W. C., 249, *258*
French, J. W., 243, *258*
Frenk, S., 228, *258*
Fuller, J. L., 83, 101, *104*
Furth, H. G., 211, *212*

G

Gaito, J., 221, *258*
Gardner, B. T., 204, *212*

Gardner, F., 242, 243, *255*
Gardner, L. E., 224, 226, *258*
Gardner, R. A., 204, *212*
Gee, W., 246, *258*
Gegenbaur, C., 135, *141*
Geils, H. D., 101, *104*
Gelber, B., 241, 242, *258*
Geoffroy Saint-Hilaire, E., 131, *141*
Geoffroy Saint-Hilare, I., 131, *141*
Geschwind, N., 204, *212*
Ghiselin, M. T., 129, 132, 133, 134, 136, 137, 139, *141*, 143, 144, 145, *150*, 155, *166*, 181, *198*
Giolli, R. A., 101, *104*
Gleitman, H., 211, *212*
Gleitman, L. R., 211, *212*
Glickman, S. E., 109, *112*
Goin, C. J., 156, 157, *166*
Goin, O. B., 156, 157, *166*
Gooch, J. L., 82, *104*
Goodrich, E. W., 136, *141*
Gottesman, I. I., 83, *104*
Gould, S. J., 117, 120, 121, *122*, 176, 178, *179*
Gourevitch, G., 7, *10*
Grabowski, U., 242, *258*
Graeber, R. C., 4, *10*
Graziadei, J., 219, *262*
Green, M. C., 101, *106*
Greenwood, P. H., 103, *104*, 123, 125, 126, *126*, *127*
Gregory, W. K., 177, *179*
Grinnell, A. D., 27, *43*
Grooves, P. M., 237, *258*
Guillery, R. W., 101, *104*
Gulick, W. L., 163, *166*
Gygax, P. A., 149, *151*

H

Haas, O., 136, *141*
Haggerty, R., 250, *257*
Haigler, H. J., 235, *258*
Hailman, J. P., 18, 19, 20, *22*, 184, 192, 193, 194, 197, *198*
Haldane, J. B. S., 48, *76*
Hall, W. C., 109, 110, *112*, 150, *150*
Hanzel, T. E., 243, *258*
Harden, C. M., 224, 225, 243, *258*
Harman, P. J., 23, *43*
Harris, J. E., 119, *122*
Hartman, H. B., 101, *105*
Hayhow, W. R., 109, *112*
Hazlett, E. A., 251, 254, *258*
Heath, J. E., 39, *43*
Hebb, D. O., 23, 25, 38, *44*
Hecht, M., 165, *167*
Heffner, H., 100, *104*
Hegmann, J. P., 101, *104*, *105*
Held, R. M., 217, *258*
Herrick, C. J., 3, *10*, 147, 148, *150*
Hershberger, W. A., 249, *258*
Herz, M. J., 247, *261*, *263*
Hess, W. R., 240, *258*
Hetherington, K., 246, *257*
Higgins, J. V., 97, *105*
Hinde, R. A., 137, *141*, 159, 164, *166*, 197, *198*
Hirsch, H. V. B., 218, 254, *258*
His, W., 177, *179*
Hodos, W., 108, 109, 111, *112*, 134, 137, *140*, *141*, 143, 144, 145, 147, *150*, 153, 154, 162, 165, *166*, 181, 185, 186, 187, 189, 191, *197*
Holloway, R. J., Jr., 71, *76*
Holmgren, B., 228, *258*
Horn, G., 228, 249, *258*, *262*
Horowitz, G., 100, *106*
Horridge, G. A., 226, 231, 232, 233, 234, 238, 250, *256*, *258*
Houghton, J. C. W., 15, *22*, 184, *198*
Hoyle, G., 232, 233, 238, 250, *259*
Hubel, D. H., 239, *259*
Huber, G. C., 3, *10*, 156, *166*
Hunsaker, D., 156, *166*
Hunter, W. S., 249, *259*
Huxley, J. S., 117, 118, 119, 120, *122*
Hyden, H., 240, *259*

J

Jacobson, A. L., 238, 244, 245, *259*, *260*
Jacobson, R., 238, *263*
Jaeger, R. G., 18, 19, *22*
Jane, J. A., 4, *10*
Jardine, N., 136, *141*
Jarvik, L. F., 93, *104*
Jensen, A. R., 93, *105*
Jensen, D. D., 222, 241, *259*
Jerison, H. J., 28, 29, 31, 32, 33, 34, 35, 36, 38, 39, 40, 41, 42, 43, *44*, 50, 51, 52, 59, 64, 69, 70, 71, 72, 73, *76*, 120, *122*
John, E. R., 23, *44*, 218, 219, 220, 233, 234, 235, 236, 245, *257*, *259*
Johnson, R. F., 103, *105*

Johnson, R. N., 162, *166*
Jones, F. G. W., 249, *262*

K

Kaas, J. H., 101, *104*
Kabat, L., 238, *263*
Källén, B., 147, *150*, 153, *166*
Kakalewski, J. W., 39, *45*
Kandel, E. R., 25, 44, 219, 229, 235, 252, *256, 259, 261*
Karten, H. J., 143, 150, *150, 151*
Kater, S. B., 101, *105*
Katz, H. M., 162, *167*
Katz, M., 241, *259*
Kaufman, D. W., 82, *106*
Kelly, D. E., 53, *76*
Kelly, S., 244, *257*
Kennedy, D., 23, 25, *44*, 227, 228, 231, 235, *256, 259, 263*
Kerkut, G. A., 240, 250, *259*
Kermack, K. A., 64, 67, *76*
Kiernan, J. A., 108, *112*
Kieso, R. A., 101, *105*
Killam, K. F., 236, *259*
Kimble, D. P., 245, *260*
Kinastowski, W., 224, *259*
Kitts, D., 52, *76*
Klopfer, P. H., 184, *198*
Krasne, F. B., 228, 246, 250, 251, *259*
Kristan, W. B., Jr., 236, *260*
Kruuk, H., 15, *22*, 184, *198*
Kuenzer, P. P., 226, 246, *260*
Kuhlenbeck, H., 147, *151*, 188, *198*
Kupfermann, I., 229, *256, 261*
Kurtén, B., 48, *76*

L

Lahue, R., 221, 222, 223, 224, 230, 235, 237, 239, 248, 253, *257, 260*
Lamarck, J. P. B., 117, *122*
Lankester, E. R., 144, *151*, 190, *198*
Lapan, E., 224, *255*
Lartet, E., 24, *44*
Lashley, D. S., 25, *44*
Lashley, K. S., 23, *44*, 219, 220, 234, *260*
Latimer, H. B., 30, *44*
Lay, D. M., 9, *10*
LeBoeuf, B. J., 98, *105*
Lee, R. M., 245, 252, *260*
Lenneberg, E. H., 199, 204, 206, 211, *212*
Lerner, I. M., 89, 91, *105*
Lettvin, J. Y., 219, 239, *260*
Levi-Montalcini, R., 232, *260*
Lickey, M. E., 252, *260*
Lindzey, G., 83, 96, *105*
Loehlin, J., 83, *105*
Long, B. S., 211, *212*
Longo, N., 249, *260*
Lorenz, K., 98, 99, *105*, 164, *166*
Luco, J. V., 234, *260*
Lund, R. D., 101, *105*
Luria, A. R., 220, *260*

M

MacIntyre, G. T., 55, *76*
Mackintosh, N. J., 251, *260*
MacLeay, W. S., 131, *141*
Magoun, H. W., 234, *261*
Makous, W. L., 248, *260*
Manosevitz, M., 83, *105*
Marquis, D. G., 26, *44*
Marsh, O. C., 24, 37, *44*
Martin, R. D., 109, *113*
Masterton, B., 100, *104*
Maturana, H. R., 219, 239, *260*
Maynard, R., 225, *262*
Mayr, E., 19, *22*, 80, 83, 92, 102, 103, *105*, 153, 162, 163, *166*
Mazochin-Porshnyakov, G. A., 250, *260*
McClearn, G. E., 83, 85, 90, 93, 94, 96, 101, *104, 105, 106*
McConnell, J. V., 238, 244, 245, *260, 263*
McCulloch, W. S., 219, 239, *260*
McGaugh, J. L., 252, *262*
McKenna, M. C., 109, *112*
Mekler, L. B., 219, *260*
Merrell, D. J., 81, *105*
Miles, R. S., 123, 125, *126, 127*
Miller, K. R., 246, *261*
Minami, H., 249, *260*
Mishkin, D., 219, *262*
Moore, R. Y., 108, *113*
Morowitz, H. J., 224, 232, *255*
Morrell, R., 235, *260*
Morrow, J. E., 251, *261*
Moruzzi, G., 234, *261*
Moskowitz, N., 26, *44*
Moss, M. L., 169, *173*
Moushegian, G., 5, *10*
Moy-Thomas, J. A., 123, *127*

Murphy, E. H., 254, *261*
Murphy, R. M., 249, *261*
Murray, J., 102, *105*
Myers, G. S., 126, *126*
Myers, R. D., 39, *44*

N

Naef, A., 136, *141*
Nauta, W. J. H., 149, 150, *151*
Neel, J. V., 96, 97, *105, 106*
Nelson, G. J., 165, *167*
Nelson, J. B., 19, *22*
Nelson, M. C., *261*
Nieuwenhuys, R., 143, 147, 148, 149, *151*
Niklaus, R., 249, *261*
Noback, C. R., 26, *44*
Noirot, E., 162, *167*
Norman, J. R., 123, *127*
Northcutt, R. G., 4, *10*, 109, *112*, 143, *151*, 153, *167*
Nyberg, D., 39, *44*, 65, *76*

O

Ohno, S., 50, *76*
Oken, L., 131, *141*
Olds, J., 238, *261*
Oliver, G. W. C., 240, 250, *259*
Olson, E. C., 52, 58, 60, 64, *76, 77*, 164, *167*
Osborn, H. F., 30, *44*
Owen, K., 83, *106*
Owen, R., 131, 132, 133, 134, 135, 136, 137, *141,* 181, *198*

P

Packard, A., 176, *179*
Palka, J. 221, 222, 239, *255, 261*
Papez, J. W., 3, *10*
Patterson, B., 64, *77*
Patterson, C., 125, *126*
Patten, D. G., 248, *263*
Peeke, H. V. S., 221, 222, 246, *255, 261, 263*
Peretz, B., 243, *257*
Peters, H., 239, *261*
Peterson, R. S., 98, *105*
Pieratoni, R., 249, *256*
Pilbeam, D., 51, *77*
Pinsker, H. M., 229, 252, *256, 261*
Pitts, W. H., 219, 239, *260*
Plavilstchikov, N. N., 238, *261*
Polikarpov, G. G., 243, *263*
Polyak, S., 40, *44*
Poole, D. F. G., 169, *173*
Premack, D., 204, *212*
Pribram, K., 162, *167*
Prosser, C. L., 156, 159, *167*
Pumphrey, R. J., 236, *261*

Q

Quiring, D. P., 28, 29, *43*

R

Radinsky, L., 51, *77*, 109, *113*, 146, *151*
Ralls, K., 100, *105*
Ranney Mize, R., 254, *261*
Rasch, E., 217, *261*
Ratner, S. C., 226, 244, 246, *257, 261*
Raup, D. M., 178, *179*
Rawdon-Smith, A. F., 236, *261*
Ray, A. J., 247, *261*
Razran, E., 235, *261*
Reed, E. W., 97, *105*
Reed, S. C., 97, *105*
Reeder, W. G., 5, 6, *10*
Remane, A., 136, *141*
Remler, M. P., 231, *259*
Rensch, B., 117, *122*
Richards, O. W., 83, *106*
Riccio, D., 244, 245, *257*
Rick, J. T., 240, 250, *259*
Riesen, A. H., 217, *261*
Roberts, A., 228, *259*
Roberts, M. B. V., 226, *261*
Roberts, R. C., 86, 90, *106*
Robinson, R., 130, *141*
Robson, G. C., 83, *106*
Roderick, T. H., 101, *106*
Roe, A., 83, *106,* 199, *212*
Romer, A. S., 3, *10,* 49, 75, *77,* 163, *167*
Rosen, D. E., 126, *126*
Rosen, J. J., 219, *262*
Rosenzweig, M. R., 217, *261*
Ross, D. M., 244, *261*
Ross, J. R., 206, *213*
Rovainen, C. M., 25, *44*
Rowell, C. H. F., 228, 249, *258, 261, 262*
Rowell, T. E., 162, *167*
Ruch, T. C., 26, *45*

Rucker, W., 243, *258*
Rupert, A. L., 5, *10*
Rushforth, N. B., 225, 243, 244, *262*
Rusinov, V. S., 235, *262*
Russell, D. E., 34, 35, *45*

S

Sacher, G. A., 37, *45*
Sanders, G. D., 251, *262*
Schaeffer, B., 123, *127,* 165, *167,* 171, 172, *173*
Schapiro, S., 217, *262*
Schietti-Cavazza, B., 249, *256*
Schlapfer, W. T., 240, *263*
Schlesinger, I. M., 211, *213*
Schopf, T. J. M., 82, *104*
Schull, W. J., 83, 96, 97, *104, 105, 106*
Scott, S., 217, *255*
Seidenstein, S., 41, *45*, 68, *77*
Selander, R. K., 82, 101, 103, *105, 106*
Selverston, A. I., 228, 231, *259, 263*
Severtsov, A. N., 117, *122*
Shannon, C. E., 191, *198*
Sheiman, E. M., 249, *256*
Sholl, D. A., 38, *45*
Shotwell, J. A., 5, 6, *10*
Sidman, R. L., 101, *106*
Sigogneau, D., 34, 35, *45*
Simmons, E. L., 51, *77*
Simpson, G. G., 48, 55, 58, 64, *77,* 83, *106,* 130, 133, 134, 136, 140, *141,* 144, 145, *151,* 162, *167,* 187, 188, *198,* 199, *212*
Sinclair, H., 211, *213*
Sivian, L. J., 7, *10*
Skeen, L. C., 245, *257*
Slobin, D. I., 211, *213*
Slotnick, B. M., 162, *167*
Sluckin, W., 218, *262*
Smith, G. E., 246, *262*
Smith, H. M., 135, *141,* 145, *151,* 188, 191, *198*
Smith, J. C., 248, *262*
Smith, M. P., 249, *258*
Smithson, E. L., 251, *261*
Sokal, R. R., 222, *262*
Spencer, W. A., 219, 222, 229, 230, 236, 246, *259, 262*
Sroges, R. W., 109, *112*
Stafekhina, V. S., 249, *256*
Stebbins, G. L., 27, 37, *45*
Stebbins, R. C., 53, *77*
Stein, D. G., 219, *262*
Stephan, H., 23, 28, 30, *43, 45*, 176, *178*
Stephens, G. J., 252, *262*
Sterba, G., 126, *127*
Steyn, W., 53, *77*
Stockton, M. D., 100, *106*
Strother, W. F., 8, *11*
Swainson, W., 132, *142*
Swift, H., 217, *261*
Szlep, R., 234, *262*
Szulc, E., 15, *22*, 184, *198*

T

Tatarinov, L. P., 58, 63, *77*
Tauc, L., 235, *259*
Tavolga, W. N., 156, *167*
Thiessen, D. D., 83, 99, *105, 106*
Thompson, D. W., 119, *121*, 178, *179*
Thompson, R. F., 222, 229, 230, 236, 237, 246, *258, 262*
Thompson, W. R., 83, *104*
Thorpe, W. H., 249, 254, *262*
Tilney, F., 25, *45*
Tilson, E. F., 100, *106*
Tinbergen, N., 15, *22,* 137, *141,* 160, 161, 164, *166, 167,* 184, 195, *198*
Tinker, S., 126, *126*
Tryon, R. C., 162, *167*
Tumarkin, A., 61, *77*

V

Valenstein, E. S., 39, *45*
Vandenberg, S. G., 83, 93, 96, *104, 106*
Van Oeveren, R., 100, *104*
Van Oye, P., 245, *262*
Van Valen, L., 109, *113*
Veale, W. L., 39, *44*
Vernberg, F. J., 126, *127*
Verprintsev, B. N., 240, *256*
Von Baumgraten, R. J. 235, *258*
Von Burg, R., 224, 234, 237, 241, 243, 248, *257*
von Frisch, K., 250, *262*
Vowles, D. M., 231, 232, 249, *262, 263*
Vukovich, K. R., 217, *262*

W

Waddington, C. H., 201, *213*
Wahlsten, D., 100, 101, *106,* 249, *258*

Walker, R. J., 240, 250, *259*
Wallace, G. J., 157, *167*
Warren, J. M., 108, *113*
Warton, S. S., 240, *256*
Wasserman, G. S., 248, *263*
Wawrzynczyk, S., 224, *263*
Weaver, W., 191, *198*
Webster, D. B., 5, 8, *10 11*
Webster, M., 5, 8, *11*
Weiskrantz, L., 24, 26, *45*
Welker, W. I., 41, *45*, 68, *77*
Wells, M. J., 251, 252, *263*
Wells, P. H., 245, 250, *263*
Welty, J. C., 158, 160, 163, *167*
Westoll, T. S., 172, *173*
Wever, E. G., 27, *45*, 156, *167*
White, J. E., 101, *105*
White, S. D., 7, *10*
Whitney, G., 83, 94, 100, *106*
Whitsett, M., 83, *106*
Whittaker, R. H., 120, *122*
Wickelgren, B. C., 229, 230, *263*
Wickler, W., 137, *142*, 159, *167*
Wiersma, C. A. G., 239, *263*
Wiesel, T. N., 239, *259*
Wietzman, S. H., 126, *126*
Willows, A. O. D., 219, 221, 222, 231, 239, 255, *255*, *257*, *263*
Wilson, D. L., 240, *263*
Wilson, D. M., 25, *45*
Wimer, C. C., 101, *106*
Wimer, R. E., 101, *106*
Wine, J. J., 231, *263*
Wirz, K., 30, *45*
Withers, R. F. J., 136, *142*
Wittgenstein, L., 130, *142*
Wood, A. E., 5, 6, *11*
Wood, D. C., 224, 225, 243, *263*
Woodsmall, K. S., 228, *259*
Woodson, F. B. J., 240, *263*
Woolsey, C. N., 23, *45*
Wyers, E. J., 246, 261, *263*
Wynne-Edwards, V. C., 98, *106*

Y

Yanagisawa, K., 239, *263*
Yang, S. Y., 101, *106*
Youniss, J., 211, *212*

Z

Zelman, A., 238, *263*
Zubkov, A., 243, *263*
Zucker, R. S., 228, *263*

Subject Index

A

Ablation technique, 219–220
Adaptation for predator avoidance in
  rodents, 5, 8
Adaptive unit, 37
Aerobacter aerogenes, 241
Agelena labyrinthica, 239
Allometry, study of, 117
Amblotherium, 64
Amia, 125
Amphioxus, 170, 173
Anagenesis, 117–119
Analog, the, 175–179
Analogy, 129–142, 181–182
  coincidental, 196
  convergent, 195
  definitions of, 138–140, 145
  iconic, 196
  kinds of, 195–196
  parallel, 195
Animal learning, 99–100
Annelids, learning in, 245–246
Anolis, 156
Antimeric homology, 139
Aplysia, 235, 236, 252
Aranea diademata, 234
Araneus diadematus, 248
Arctocyon, 30, 35, 36
Arctocyon primaevus, 34, 35
Archaeopteryx, 51, 183
Arrowpoison frogs, 18
Arthropods, learning in, 247–251
Articulata, neural characteristics of, 2
Auditory system
  in rodents, 5–10
  sensitivity, 5, 7–9
  transformer ratio, 5, 7
Australopithecus africanus, 69
Australopithecus habilis, 69

B

Barylamoda, 31
Basic dyes, see Cresyl violet; Thionin
Behavior
  causes and origins, 13–17
    biological function, 15
    dynamic control, 14
    ontogenetic development, 14–15
    phylogenetic history, 15–16
  comparative method, 13–22
  effects on niche or adaptive zone, 83
  evolution of, 153–167
  image driven, 239
  role in evolution, 82–85
  structural correlates, 155
  as taxonomic character, 163–164
Behavioral homology
  definition of, 156
  limitations of, 164–165
Behavioral homoplasy, 158–161
Biochemical correlates of learning,
  240–241
Blattela gumanica, 249
Blatta orientalis, 234

Brain–body relations, 28–30
Brain size, assessing behavior, 69–74
Buteo albonotatus, 159

C

Calamoichthys, 125
Canis latrans, 36
Carassius auratus, 98
Chamaeleo, 156
Chlamydoselache, 124
Chance similarity, definition of, 144
Chondrichthyes, 123–124
Cladogenesis, 117–119
Coelenterates, learning in, 243–244
Coincidental analogy, 140, 196
Common cncestry, 183–184
Comparative method, 1–22
  in ethology, 17–21
  history of, 1–3
  rules of, 2
  uses in behavioral studies, 13–22
Comparative neurology, 107–113
  ancestral and dervied characteristics, 109–112
Conservation principal, 37–40
Convergence, 159
Convergent analogy, 139, 195
Convergence, definition of, 144
Corticalization of function, see encephalization
Cosyphodon hamatus, 31
Cranial capacity, see brain size
Cresyl violet stain, 148
Critical periods, 254
Cynodont, shift in masticatory mechanism, 61

D

Diadectes, 62
Didelphis marsupialis, 28
Dimetrodon, 117
Dipodomys, 5, 7, 8, 9, 10, 11
  Dipodomys merriami, 7
Domestication, use in posing evolutionary questions, 97–102
Drosophila, 90

E

Edaphosaurus, 117
Encephalization, 24, 26–43
Encephalization (*contd.*)
  forebrain and hindbrain, 30–34
  olfactory bulbs, 34–37
  quotient, 41–42
Endothermy, evolution of, 39
Eozostrodon, 64
Epihippus, 69
Epistatic effects, 87
Erinaceus europaeus, 28
Ethology, comparative method in, 17–21
Evolution
    adaptive unit, 37
    behavioral, 153–167
    comparative neurology, 107–113
    ancestral and derived characteristics, 109–112
    conservation principal, 37–40
    domestication as model 97–102
    encephalization quotient, 41–42
    endothermy, 39
    genetic considerations, 79–103
      natural populations, 80–82
      population approach, 79–80
    intelligence, 92–97
    mosaic, 164
    principal of adaptiveness, 41
    quantum, 55
    rates of, 47–75
      nervous system behavior, 51–54
      sensory systems, 53–54
    role of behavior, 82–85
    selection pressures, 40–42
    sensory, chart, 56–57
    use of grades and clades, 115–122
  behavioral 49–51
    brain, true mammals, 66–74
    chart, 56–57
    pelycosaurs to therapsides, 58–64
    quasi-mammals, 64–66
Evolutionary homology, 139

F

Feedback circuitry, 232
Fiber connections, usefulness in determining CNS homologies, 149–150
Field homology, 158
Fishes, availability and phylogeny of, 123–127
Forebrain, encephalization, 30–34
Functional neuroanatomy
  ablation technique, 219–220

Functional neuroanatomy (*contd.*)
equipotentiality, 220
pluripotentialism, 220–221

G

Gallus domesticus, 98
Genetics
causal mechanisms of evolution, 79–103
dominance, 86–87
epistatic effects, 87
evolution, causal mechanisms of, 79–103
population approach, 79–80
heterosis, 90–91
heritability, 87–88
inbreeding depression, 90–91
and intelligence, 92–97
of natural populations, 80–82
polymorphisms, 81
quantitative population model, 85–88
selections, 88–102
stabilization of, 91–92
Golgi method, 147
Grapic double integration, 32, 35, *see also* Brain–body relations

H

Habituation, 236–237
Helix aspersa, 252
Heritability, 87–88
Heterosymaptic facilitation, 235–236
Hindbrain encephalization, 30–34
Homo sapiens, 204
Homogeny, 185–189
definition of, 144, 187
Homology, 129–142
basis of, 155–156
behavioral, 153–167
definitions of, 156
limitation of, 164–165
morphological considerations, 84
special cases, 157–158
commonalities of characters, 182–185
appearance, 185
ancestry, 183–184
functions, 184–185
definitions of, 138, 144, 185–189
criteria for, 189–192
field, 158
iterative, 195
kinds of, 195–196

Homology (*contd.*)
latent, 171
logic analysis, 181–198
use of Venn diagrams, 186–188
nervous system, 143–151
kinds of data, 147
usefulness of fiber connections, 149–150
mammalian and nonmammalian homologies, 148
practical aspects, 169–173
Homoplasy, 182, 185–189
behavioral, 158–161
definition of, 144
Hybrid vigor, *see* Heterosis
Hydra, 242, 244
Hydrolagus, 124
Hyopsodus miticulus, 31
Hyperexcitability, nervous system, 234–236
Hyracotherium, 69

I

Iconic analogy, 140
Image-driven behavior, 239
Inbreeding depression, 9–91
Information processing, 232–233
Integration, 232–233
Intelligence, invertebrate, 215–263
invertebrate–vertebrate differences, 253–255
strategies of inquiry, 217–233
changes in system components, 218
comparative, 221–233
epigenetic, 217–218
learning, 218–221
preprogrammed strategies, 217
Invertebrate intelligence, 215–263
Invertebrate neural mechanisms, 231–233
cell morphology, 231
feedback circuitry, 232
glial presence, 232
integration and information processing, 232–233
nerve process differentiation, 231–232
Iterative homology, 139, 195

K

Kangaroo mice, see *Microdipodops*
Kangaroo rats, see *Dipodomys*
Kittiwake gulls, 19

L

Language
and brain, 204
and children, 210–212
compatative study, 199–213
Latent homology, 171
Latimeria, 125
Learning, animal, 99–100
in annelids, 245–247
in arthropods, 247–251
biochemixal correlates, 240–241
in chelicreates, 247–248
in coelenterates, 243–244
in crustacea, 250–251
in insects, 248–250
invertebrate, 215–263
capacities, 233–252
in mollusks, 251, 252
in planarians, 244–245
in protozoans, 241–243
reversal, 223
strategies of inquiry, 217, 233
changes in system components, 218
comparative, 221–233
learning mechanisms, 218–221
Lepidosirent, 125
Lepisosteus, 125, 135
Limulus, 234, 237, 240, 248
Limulus polyphemus, 223
Liomys, 5, 7
Localization of function, 24–25

M

Marchi method, 147
Mass action doctrine, 25
Memory transfer, 237–238
Meniscotherium robustum, 31
Mesophippus, 69
Misozoic ceratodus, 125
Microdipodops, 5, 7, 8, 9
Mimicry, 159–160, 196, *see also* Iconic analogy
Mimus, polyglottos, 160
Mirounga augustirostris, 98
neural characteristics, 2
Mollusks, 251
learning in, 251–252
Morganucodon, 64
Mosaic evolution, 164
Mus musculus, 91, 100

N

Nauphoeta concera, 249
Nautilus, 131
Nemeritis canescens, 249
Neoceratodus, 125
Nervous system, morphological homology, 84
kinds of data, 147
mammalian and nonmammalian homologies, 148
usefulness of fiber connections, 149–150
Neurohistological methods, *see* Cresyl violet stain; Golgi method; Marchi method; Thionin stain
Neural mechanisms, invertebrate–vertebrate differences, 231–233
feedback circuitry, 232
glial presence, 232
integration and information processing, 232–233
cell morphology, 231
nerve process differentiation 231–232
neuron numbers, 232
Niche changes initiated by behavior, 83

O

Olfactory bulbs, encephalization of, 34–37
Ontogenetic development of behavior, 14–15
Orohippus, 69
Osteichthyes, 124–125

P

Palaeontologia Africana, 75
Pantolambda bathmodon, 31
Parallel analogy, 139, 195
Parallelism, definition of, 144
Paramecia, 241
Paramecium Aurelia, 241
Passer domesticus, 103
Passerella iliaca, 16, 184
Pelycasaurs, brain evolution, 58–59
Periplaneta americana, 249
Perognathus, 5, 7, 8, 9
Petromyzon, 176
Phenocodus, 30
Phenacodus primaevus, 31, 32, 34
Photoreceptors, cones in amphibians, 18

Phylogenetic history, of behavior, 15–16
Pocket mice, see *Heteromys; Liomys; Perognathus*
Polypterus, 125, 143, 148
Porcellio, 251
Postilla, 75
Posttetanic potentiation, 235–236
Principal of adaptiveness, 41
Principal of proper mass, 24, 25
Procambarus clarkii, 232
Protopterus, 125
Protozoa, learning in, 241–243
Ptilodus, 28, 64

Q

Quasi-mammals, brain evolution in, 64–66

R

Radiata neural characteristics, 3
Ramapithecus, 69
Rangifer tarandus, 99
and *esculenta*, 235
Rattus norwegicus, 98
Reticular formation, 146
Reversal learning, 223
Ritualization, of behavior, 195
Rissa tridactyla, 159
Rodents
auditory system, 5–10
correlations among species, 9
habitats, 5
phylogenetic history of Heteromyidae, 6
sensitivity, 5, 7–9

S

Salmo salar, 130
Scala natural, 2
Sceloporus, 156
Selection pressures, 40–42
Selective advantage of behavior, 15
Sensory evolution, chart, 56–57
Sensory projections in vertebrates, 2–4
Serial homology, 139
Sexual homology, 139
Sinoconodon, 64
Stentor, 243
Stephanoaetos coronatus, 160
Stomphia coccina, 244
Sula bassana, 159
Swallowtailed gulls, 19

T

Taxanomy, 18–19
Tenebrio molitor, 249
Therapsids, brain evolution, 58–64
Thionin stain, 148
Tetrahymena pyriformis, 242
Torostoma rufum, 160
Toxostoma, 16
Toxostoma rufun, 194
Triconodon, 28, 29
Tritonia, 222

U

Uintatherium anceps, 31, 37
Unit specifications, 239–240
Unit conditioning, 238–239

V

Vertebrate neural mechanisms, 231–233
cell morphology, 231
feedback circuitry, 232
glial presence, 232
integration and information processing, 232–233
nerve process differentiation, 231–232
Vertebrate neural characteristics, 2–4
Visual system pathways in primate lineage, 109–110